STRUCTURE AND BONDING is issued at irregular intervals, according to the material received. With the acceptance for publication of a manuscript, copyright of all countries is vested exclusively in the publisher. Only papers not previously published elsewhere should be submitted. Likewise, the author guarantees against subsequent publication elsewhere. The text should be as clear and concise as possible, the manuscript written on one side of the paper only. Illustrations should be limited to those actually necessary.

Manuscripts will be accepted by the editors:

Professor Dr. *J. D. Dunitz*   Laboratorium für Organische Chemie der Eidgenössischen Hochschule
CH-8006 Zürich, Universitätsstraße 6/8

Professor Dr. *P. Hemmerich* Universität Konstanz, Fachbereich Biologie
D-7750 Konstanz, Postfach 733

Professor *J. A. Ibers*   Department of Chemistry, Northwestern University
Evanston, Illinois 60201/USA

Professor Dr. *C. K. Jørgensen* 51, Route de Frontenex,
CH-1207 Genève

Professor *J. B. Neilands*   University of California, Biochemistry Department
Berkeley, California 94720/USA

Sir *Ronald S. Nyholm*, FRS †

Professor Dr. *D. Reinen*   Institut für Anorganische Chemie der Universität Marburg
D-3550 Marburg, Gutenbergstraße 18

Professor *R. J. P. Williams* Wadham College, Inorganic Chemistry Laboratory
Oxford OX1 3QR/Great Britain

SPRINGER-VERLAG

D-6900 Heidelberg 1
P. O. Box 1780
Telephone (06221) 4 91 01
Telex 04-61 723

D-1000 Berlin 33
Heidelberger Platz 3
· Telephone (0311) 82 20 01
Telex 01-83319

SPRINGER-VERLAG
NEW YORK INC.

175, Fifth Avenue
New York, N. Y. 10010
Telephone 673-2660

STRUCTURE AND BONDING

Volume 14

Editors: J. D. Dunitz, Zürich
P. Hemmerich, Konstanz · J. A. Ibers, Evanston
C. K. Jørgensen, Genève · J. B. Neilands, Berkeley
Sir Ronald S. Nyholm†, London · D. Reinen, Marburg · R. J. P. Williams, Oxford

With 52 Figures

Springer-Verlag
Berlin Heidelberg GmbH 1973

ISBN 978-3-540-06162-5   ISBN 978-3-540-38371-0 (eBook)
DOI 10.1007/978-3-540-38371-0

Library of Congress Catalog Card Number 67-11280.

Contents

Structural Chemistry of Polynuclear Transition Metal Cyanides

Andreas Ludi and Hans Ulrich Güdel*

Institut für anorganische Chemie, Universität Bern, CH-3000 Bern 9, Switzerland

Table of Contents

I. Introduction

Prussian blue, $Fe_4[Fe(CN)_6]_3 \cdot xH_2O$, the first recorded synthetic coordination compound and the prototype of the polymeric cyanides, has been the object of a great number of investigations and speculations (a survey of the literature is given in Ref. 1). A wide variety of physical methods has been used to elucidate the structure and bonding of this compound. For a long time the central question was whether the struc turally different iron ions can be associated with distinct valences or

* Present address: Research School of Chemistry, Australian National University, Canberra, Australia.

whether they oscillate between different valences. The wealth of information accumulated today from X-ray diffraction data (2—4), infrared spectroscopy (5, 6), magnetic susceptibility measurements (5, 7), electronic spectroscopy (8), Mössbauer studies (9—13) and photoelectron spectroscopy (14, 15) is consistent with the formulation of Prussian blue as the iron(III) salt of hexacyanoferrate(II). The presence of potassium or other alkali metals in various amounts does not affect the distinct oxidation states of the different iron ions. Prussian blue and its analogs for numerous metal ions are usually obtained as extremely fine powders or even as gels (16). Whereas the basic features of the crystal structures of these generally cubic polymers have been known since the pioneering work of *Keggin* and *Miles* (2), the complete characterization of crystal structures has not been accomplished until recently when suitable single crystals could be prepared.

The attraction that the polynuclear transition metal cyanides have always had for chemists may be ascribed mainly to the characteristic deep colors of some members of this class of compounds. These inorganic polymers offer a rich variation in their stoichiometries and hence in their electronic properties. The broad range of different compounds having the general composition $M_k^A[M^B(CN)_m]_l \cdot xH_2O$ is due to the large number of possible combinations of transition metal ions M^A and M^B, particularly by considering the different possible oxidation states. As special cases we will find a number of mixed valence compounds, a class of compounds which has attracted widespread attention during the last few years (17, 18).

The ambident nature of the cyanide ion leads to a four-atomic sequence $M^A-N-C-M^B$, which may be connected in different ways to form a crystal lattice. The coordination behavior of the two different metal ions decides whether there will be a one-, two-, or three-dimensional linkage. Examples with typical chains are represented by the crystals of $AgCN \cdot 2\ AgNO_3$ (19) and $Ni(en)_2Pd(CN)_4$ (20). A two-dimensional linkage is seen in the sheets of the cyanide-ammonia clathrates (21, 22). The vast majority of the polynuclear cyanides, however, contain an octahedral hexacyanometalate, $M^B(CN)_6^{n-}$, and have a three-dimensional framework structure (19, 23).

Although the metal-carbon bond in cyano compounds has been confirmed by neutron diffraction in only a limited number of substances (24—29), it is generally and reliably accepted that in the stable mononuclear complexes the metal is always bonded to carbon. Support for the M—C—N linkage is provided by electronic spectra (30) and by vibrational spectroscopic data obtained with different isotopes (31).

The ambident coordination of the cyanide in the polymeric compounds is most easily demonstrated by infrared spectroscopy. The stretching frequency $\nu(CN)$ of a mononuclear complex $M^B(CN)_6^{n-}$ is increased by approximately 30–70 cm^{-1} upon forming the linkage M^B–C–N–M^A of the polynuclear cyanide (32). In most cases the formation of the poly-nuclear compound does not change the coordination environment of the originally carbon-coordinated metal ion, M^B. Only in a few polynuclear cyanides has the isomerization M^A–N–C–M^B → M^A–C–N–M^B been observed (33).

Various aspects of the transition metal cyano complexes have been discussed in three authoritative reviews (1, 16, 23). The present authors will therefore not attempt to give a comprehensive review of the subject. Instead, we restrict ourselves to the discussion of structural properties of crystalline cyano compounds containing octahedral $M^B(CN)_6$ groups, where the characteristic construction element is given by the four-atomic sequence M^A–N–C–M^B. M^B represents a transition metal, M^A either a transition metal or a metal of class B (34), or hydrogen.

II. The Structure of Prussian Blue Analogs

1. Background

Prussian blue analogs are here defined as polynuclear transition metal cyanides of the composition $M_k^A[M^B(CN)_6]_l \cdot x H_2O$ crystallizing with a cubic unit cell. They are easily obtained as sparsely soluble precipitates by mixing solutions of a cyano complex $M^B(CN)_6^{n-}$ with an appropriate salt of M^A. The compounds prepared by using the hexacyanometalate in the form of the most common potassium salt invariably contain different amounts of potassium, which in some cases can be exchanged by cesium (35). Very often the Prussian blue analogs have been formulated with a definite amount of potassium, e.g. $KFeFe(CN)_6$, $KFeCr(CN)_6$, K_2CuFe $(CN)_6$. The sparse published analytical data (6, 32, 36), however, indicate that potassium has to be considered as an impurity of these often colloidal precipitates. Thus, the polynuclear cyanides containing potassium or other alkali ions are non-stoichiometric compounds rather than com-pounds showing a definite formula as far as the alkali ions are concerned.

The precipitates of Prussian blue analog cyanides always contain variable amounts of water, which can be removed without significant effects on the X-ray diffraction pattern. Some of these water molecules can be replaced by other molecules such as ammonia or alcohol (37). It was therefore assumed that the water is present partly as zeolitic, partly as surface water (3). Recent infrared spectroscopic studies, however, reveal that, in addition, coordinated water molecules are also present (33, 38).

The first structural investigations of Prussian blue analogs date back to 1936 when *Keggin* and *Miles* (2) studied the X-ray powder patterns of iron cyanides. They found a unit cell of the face-centered type and deduced a structural model from these geometrical data. This description, which will be briefly outlined below, has until recently been accepted by many authors (3, 39—44) for the discussion of the structural properties of cubic polynuclear cyanides $M_k^A[M^B(CN)_6]_l \cdot xH_2O$. In many cases the physically equivalent description has been used (3, 42, 44) with the origin shifted by $\left(\frac{1}{2}, \frac{1}{2}, \frac{1}{2}\right)$ as compared with the original paper (2).

2. The Structural Model of Keggin and Miles

The unit cell shown in Fig. 1 is of the cubic face-centered type and has a cell edge of about 10 Å. This cell may be thought of as divided into eight small cubes which will hereafter be referred to as octants. According to the model, the positions 4a (0, 0, 0) and 4b $\left(\frac{1}{2}, \frac{1}{2}, \frac{1}{2}\right)$ (45) of the cubic face-entered unit cell are occupied by the metal ions M^A and M^B, respectively. Carbon and nitrogen atoms are situated on two sets of the 24-fold position (x, o, o). The two fourfold positions, $\left(\frac{1}{4}, \frac{1}{4}, \frac{1}{4}\right)$ and $\left(\frac{3}{4}, \frac{3}{4}, \frac{3}{4}\right)$, are randomly occupied by the required number of M^A or potassium ions in order to account for the stoichiometry. Water molecules are assumed to sit in the cavities of the lattice, but they are not assigned to defined crystallographic positions.

According to this description, the basic feature of the structure is a three-dimensional connection of M^BC_6 and M^AN_6 octahedra. Whereas all the ions M^B are in a well defined coordination environment, the unit cell contains two different kinds of metal ions M^A.

The ions of the first kind are in the octahedral nitrogen holes (position 4a) of the polymeric framework, whereas the ions of the second kind are assumed to be present as uncoordinated interstitial ions at the centers of the eight small cubes. Due to the postulated random distribution of these interstitial M^A ions, the sites $\left(\frac{1}{4}, \frac{1}{4}, \frac{1}{4}\right)$ and $\left(\frac{3}{4}, \frac{3}{4}, \frac{3}{4}\right)$ are equivalent. The structural model is thus centrosymmetric, and it has been referred to in terms of space group O_h^5-Fm3m. Although the symmetry elements of the less symmetrical space groups T_h^3-Fm3 and O^3-F432 are sufficient for the description, we will follow the convention and use O_h^5-Fm3m.

For the four most frequently realized stoichiometries of Prussian blue analogs $M_k^A[M^B(CN)_6]_l \cdot xH_2O$ the model of *Keggin* and *Miles* postulates the structural properties summarized in Table 1.

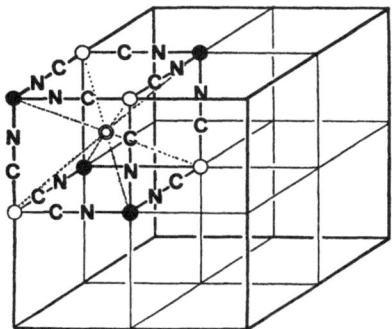

Fig. 1. The cubic unit cell of Prussian blue analogs $M_k^A[M^B(CN)_6]_l \cdot xH_2O$.
●: M^A (position 4a),　○: M^B (position 4b),　◎: position 8c (cf. text)

Table 1. *Structural properties of $M_k^A[M^B$ $(CN)_6]_l \cdot xH_2O$ with different k and l values according to the model of Keggin and Miles. Z represents the number of formula units per unit cell and n the number of M^A ions distributed at random on position 8c in space group O_h^5-Fm3m*

k	l	Z	n
1	1	4	0
4	3	$1\frac{1}{3}$	$1\frac{1}{3}$
3	2	2	2
2	1	4	4

3. The Modified Description

It cannot be expected that a structural model derived purely from X-ray powder data would provide a complete and reliable description of the actual structure. One, and probably the most important feature of the model of *Keggin* and *Miles*, however, seems to be beyond any doubt; namely the linear arrangement M^A—N—C—M^B—C—N—M^A along the edge of the unit cell. The unit cell constants of a wide variety of Prussian blue analogs have been determined. All the lattice constants measured so far are between 9.9 and 10.9 Å. Since the C—N distance is known to be close to 1.14 Å (*16*), the differences in the cell constants directly reflect the differences in the distances M^A—N and M^B—C.

A very simple but highly significant way of testing a structural hypothesis is, of course, the determination of the density to verify the

5

postulated content of a given unit cell. Until recently, the colloidal nature of the polymeric cyanides and the ill-defined amount of water and potassium made it extremely difficult to secure reliable data. Because for a long time no single crystals were available, the Keggin-Miles model could not be tested by a complete crystal-structure analysis.

It has proved possible by very slow precipitation, diffusion techniques and gel crystallization, to significantly increase the particle size of several polynuclear transition metal cyanides (38, 46). To exclude contamination with alkali metals, the hexacyanometalate was used as the corresponding acid or as the salt of a large organic cation, e.g. n-tetrabutylammonium. In some cases crystals were grown which were suitable for X-ray single-crystal studies (46—49). The systematic collection of lattice parameters, densities, analytical data, and infrared spectra prompted a modification of the original structural model proposed by *Keggin* and *Miles* (2). The principal features of this modification are as follows:

The assumption of randomly distributed interstitial, i.e. uncoordinated, metal ions M^A at position 8c is discarded. Only the positions 4a and 4b, the two fourfold positions of the cubic face-centered unit cell, are considered as possible sites for the two metal atoms. Consequently, the occupancies of these two sites depend on the composition of the cyano compound $M_k^A[M^B(CN)_6]_l \cdot x H_2O$, more specifically, on the stoichiometric ratio $k:l$. Positions 4a and 4b are both completely occupied only when $k=l=1$. For all other stoichiometries, the positions assigned to M^B, C, and N are assumed to have occupancies less than 1, whereas position 4a is still occupied by the complete set of four M^A ions. The absence of whole $M^B(CN)_6$ octahedra leads therefore to interruptions in the very tight three-dimensional $M^A-N-C-M^B-C-N-M^A$ framework postulated by *Keggin* and *Miles*. Only for the special case $k=l=1$ can we obtain an uninterrupted framework with regular M^AN_6 octahedra. For other ratios $k:l$ the number of nitrogen atoms coordinating the metal ions M^A is less than six. The sites in the coordination octahedron of M^A not occupied by nitrogen are now assumed to be filled with oxygen atoms O_I of water molecules. Evidence for the presence of coordinated water has been found in the infrared spectra of Prussian blue analogs (33, 38). In order to account for the analytically determined water content (38, 46), the presence of a second kind of water (O_{II}) is postulated near the position $\left(\frac{1}{4}, \frac{1}{4}, \frac{1}{4}\right)$, i.e. near the centers of the octants of the unit cell. This latter type of water is of zeolitic nature.

The results obtained by applying these general principles to structures with different stoichiometries are summarized in Table 2. A comparison with Table 1 shows that the original model and the modified one predict identical results for the stoichiometry $k = l = 1$. With increasing ratio $k : l$ the structural differences between the two models become more pronounced. Whereas the model of *Keggin* and *Miles* postulates increasing densities, the densities should decrease according to the modified model. A comparison of the measured and calculated densities of some twenty Prussian blue analogs gives very strong support to the modified description. A selection of lattice parameters and densities is given in Table 3.

Table 2. *Structural properties of Prussian blue analogs* $M_k^A[M^B(CN)_6]_l \cdot x H_2O$ *according to the modified model.* $p =$ *occupancy of the sites of* M^B, C, *and* N; $Z =$ *number of formula units per cell;* $y =$ *number of coordinated water molecules per cell*

k	l	p	Z	y	coordination units (average composition)
1	1	1	4	0	M^BC_6, M^AN_6
4	3	$\frac{3}{4}$	1	6	M^BC_6, $M^AN_{4.5}O_{1.5}$
3	2	$\frac{2}{3}$	$1\frac{1}{3}$	8	M^BC_6, $M^AN_4O_2$
2	1	$\frac{1}{2}$	2	12	M^BC_6, $M^AN_3O_3$

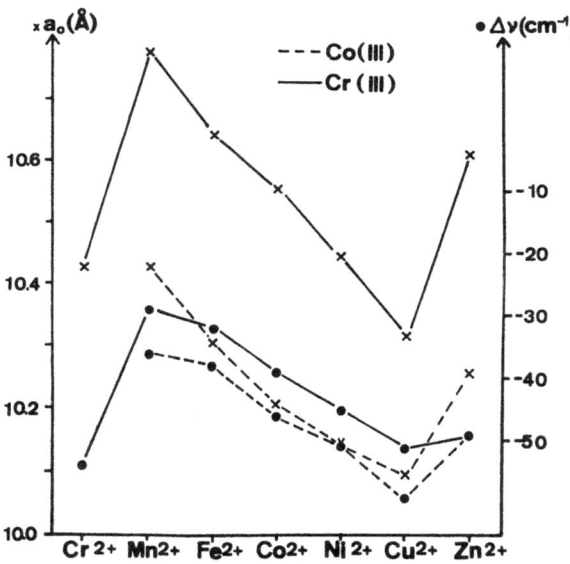

Fig. 2. Lattice constants and shifts of the stretching vibration $\bar{\nu}$ (CN) of hexacyanocobaltates(III) and -chromates(III) of divalent metals

A. Ludi and H. U. Güdel

Table 3. *Lattice constants of Prussian blue analogs* $M_k^A[M^B(CN)_6]_l \cdot x H_2O$. $Z =$ *number of formula units per cell; a = edge of the cubic unit cell (Å); d_m, d_c = measured and calculated densities, respectively (gcm^{-3})*

M^A	M^B	a	d_m	d_c
$k=1, l=1, Z=4, x=0$				
Cd(II)	Pd(IV)	10.91	1.92	1.92
$k=4, l=3, Z=1, x=13-16$				
Fe(III)	Fe(II)	10.16	1.78	1.79
Fe(III)	Ru(II)	10.26	1.85	1.94
Fe(III)	Os(II)	10.33	—	—
$k=3, l=2, Z=1\frac{1}{3}, x=11-14$				
Cr(II)	Cr(III)	10.43	1.68	1.61
Mn(II)	Cr(III)	10.77	1.50	1.48
Fe(II)	Cr(III)	10.65	1.55	1.54
Co(II)	Cr(III)	10.56	1.59	1.59
Ni(II)	Cr(III)	10.45	1.65	1.64
Cd(II)	Cr(III)	10.93	1.74	1.71
Mn(II)	Co(III)	10.42	1.65	1.62
Fe(II)	Co(III)	10.31	1.67	1.68
Co(II)	Co(III)	10.21	1.74	1.75
Ni(II)	Co(III)	10.15	1.71	1.78
Cu(II)	Co(III)	10.00	1.79	1.88
Zn(II)	Co(III)	10.26	1.75	1.77
Cd(II)	Co(III)	10.59	1.89	1.87
$k=2, l=1, Z=2, x=10-12$				
Cu(II)	Fe(II)	9.93	1.87	1.88
Cu(II)	Ru(II)	10.21	1.89	1.87
Cu(II)	Os(II)	10.23	2.18	2.13

The validity of the modified model has been tested by several complete crystal-structure determinations. Up to the present, almost all the single-crystal studies have been carried out with compounds of the stoichiometry $M_3^A[M^B(CN)_6]_2 \cdot x H_2O$ ($M^A = Mn^{2+}$, Cd^{2+}; $M^B = Co^{3+}$, Cr^{3+}, Ir^{3+}). The compound $Cs_2LiCo(CN)_6$ studied by *Wolberg* (50) is not considered as a polynuclear cyanide in this context, M^A being neither a transition metal nor a metal of class B (34). Despite great structural similarities with the Prussian blue analogs, it is rather an alkaline salt of the mononuclear cyanocobaltate(III). This distinction is, of course, arbitrary.

8

The determination of the crystal structures was complicated by the high degree of disorder. Obviously there is a very strong correlation between the parameters of the oxygen atoms O_I and the nitrogen atoms in any least-squares refinement scheme. Both atoms statistically occupy crystallographic sites with very similar coordinates.

The results of all the single-crystal investigations carried out so far for compounds $M_3^A[M^B(CN)_6]_2 \cdot x H_2O$ are in good agreement with the postulated properties of the modified model. Position 4a (0, 0, 0) of the cubic face-centered unit cell is fully occupied by four ions M^A, and $2\frac{2}{3} M^B$ ions are situated at the fourfold position 4b $\left(\frac{1}{2}, \frac{1}{2}, \frac{1}{2}\right)$, thus giving an occupancy of 2/3. Carbon and nitrogen atoms, also with occupancies of 2/3, occupy two sets of the 24-fold position (x, o, o). Eight oxygen atoms (O_I) are situated close to the eight empty nitrogen sites and eight more oxygen atoms (O_{II}) close to the special position 8c $\left(\frac{1}{4}, \frac{1}{4}, \frac{1}{4}\right)$. The metal ions M^B are octahedrally coordinated by six carbon atoms. The coordination sphere of the metal ion M^A contains nitrogen as well as oxygen, the average composition being $M^A N_4 O_2$. All the cyanide ions act as bridges, the carbon end always pointing toward M^B. The water molecules coordinated to the metal ion M^A (O_I) are assumed to be linked to the zeolitic molecules (O_{II}) by hydrogen bonds. The shortest $O_I - O_{II}$ distances are between 2.50 and 2.90 Å, in good agreement with the corresponding distances estimated from infrared spectra $(38, 46)$.

The unit cell thus contains a total of 16 molecules of water, corresponding to the composition $M_3^A[M^B(CN)_6]_2 \cdot 12 H_2O$. The analytical data and the densities indicate a water content of 11 to 14 molecules per formula unit. The degree of hydration is very sensitive to changes in humidity and temperature. This rather broad range of hydration follows quite naturally from the zeolitic nature of part of the water. The structure factor calculations have been carried out on the basis of 12 water molecules per formula unit. Attempts to locate additional water have not been fruitful.

A summary of the results of various single-crystal studies is presented in Table 4. The differences in the corresponding bond lengths are nearly within the estimated standard deviations, which are rather large because of the limited number of reflections allowed by the high symmetry of the crystals.

The crystal-structure determinations using single crystals of the compounds listed in Table 4 are consistent with a space group belonging to Laue class O_h-m3m, i.e. O^3-F432, T_d^2-F43m or O_h^5-Fm3m (48). M^A, M^B, C, and N all occupy sites that have identical multiplicities in all the three possible space groups. Neglecting the hydrogen atoms, the space group is therefore determined only by the positions of the oxygen atoms O_I and O_{II}. These atoms are found to be statistically distributed on general positions, leading thus to a disordered structure. In the most recent investigations of $Cd_3[Cr(CN)_6]_2 \cdot x H_2O$ (48) and $Mn_3[Cr(CN)_6]_2 \cdot x H_2O$ (51), equally good agreement between the calculated and observed structure amplitudes was obtained in all three space groups. It is not therefore possible to decide whether the distribution of the oxygen atoms, and thus the whole crystal structure, is centrosymmetric or non-centrosymmetric. It can be concluded, however, that if the latter is the case, the deviation from centrosymmetry has to be small.

Table 4. *Interatomic distances of Prussian blue analogs obtained from single-crystal studies*

		M^B—C	C—N	M^A—N	M^A—O
$Mn_3[Cr(CN)_6]_2 \cdot x H_2O$	(51)	2.06	1.12	2.20	2.36
$Cd_3[Cr(CN)_6]_2 \cdot x H_2O$	(48)	2.05	1.14	2.27	2.39
$Mn_3[Co(CN)_6]_2 \cdot x H_2O$	(46)	1.86	1.14	2.21	2.39
$Cd_3[Co(CN)_6]_2$	(49)	1.91	1.13	2.25	
$Cd_3[Co(CN)_6]_2 \cdot x H_2O$	(49)	1.91	1.12	2.26	
$Cd_3[Ir(CN)_6]_2 \cdot x H_2O$	(49)	2.03	1.13	2.25	
$CdPd(CN)_6$	(49)	2.05	1.14	2.27	

Attempts were also made to constrain the oxygen atoms O_I and O_{II} to the special positions (x, o, o) and $\left(\frac{1}{4}, \frac{1}{4}, \frac{1}{4}\right)$, $\left(\frac{3}{4}, \frac{3}{4}, \frac{3}{4}\right)$, respectively. These oxygen coordinates, leading to $O_I - O_{II}$ distances of 3.5 to 4.0 Å (48), however, were found to be unrealistic. Oxygen-oxygen distances in hydrogen-bonded systems usually vary between 2.5 and 3.0 Å (52). On the basis of a statistical distribution of the two different oxygen atoms on the general position, the $O_I - O_{II}$ distances found in the structures of Table 4 are all within this range. Infrared spectra indicate the existence of O—H—O bonds (38, 46). It is therefore assumed that these hydrogen bonds connect O_I to O_{II}, and one can understand that the water molecules do not occupy the symmetric sites (x, o, o) and $\left(\frac{1}{4}, \frac{1}{4}, \frac{1}{4}\right)$, $\left(\frac{3}{4}, \frac{3}{4}, \frac{3}{4}\right)$, respectively.

In his compilation of structural data, *Wyckoff* (53) relates the structures of Prussian blue analogs to the K_2PtCl_6 type. Whereas this comparison is stoichiometrically obvious for the compounds $M_2^A M^B(CN)_6$, the ambient coordination behavior of the cyanide ligand is not considered as a structural element. The polymeric cyanide is here assumed to be composed of discrete $M^B(CN)_6$ octahedra, all the metal ions M^A being placed at the centers of the octants, thus leaving all the octahedral nitrogen holes empty. For the purpose of drawing structural relationships, a more appropriate choice is the ideal cubic perovskite structure (54) where the bifunctional oxygen is replaced by cyanide and titanium alternatively by M^A and M^B.

While the general principles of the crystal structures of Prussian blue analogs have been conclusively elucidated, there still remain problems to be solved. It would be of interest to improve the resolution of the structure analysis to obtain finer details of the bond distances, and especially to study the influence of different metal ions on the C–N distance. The most desirable goal, of course, is still to grow single crystals of the archetype of these compounds, Prussian blue.

Note added in proof: A single-crystal x-ray study of Prussian Blue has been published in Chem. Commun. 1299 (1972).

III. Structural Properties of Miscellaneous Non-Cubic Polynuclear Cyanides

Whereas within the family of the cubic Prussian blue analogs a large number of lattice constants have been determined, little attention has been devoted so far to polymeric cyanides not belonging to the cubic system. It must be emphasized, however, that polynuclear cyanides having unit cell symmetries other than cubic are by no means rare exceptions. Hexacyanometalates(III) of Zn^{2+} and Cd^{2+} are obtained not only in a cubic modification but also as samples with complicated and not yet resolved X-ray patterns of definitely lower symmetry than cubic (55). The exact conditions for obtaining either modification are not yet known in detail. The hexacyanoferrates(II), -ruthenates(II), and -osmates(II) of Mn^{2+} and several modifications of the corresponding Co^{2+} salts show very complicated X-ray powder patterns which cannot be indexed in the cubic system (55). Preliminary spectroscopic studies show the presence of nearly octahedral $M^B C_6$-units in these compounds, too.

A complete crystal structure analysis has been carried out for $Mn_2Ru(CN)_6 \cdot 8 H_2O$, the corresponding hexacyanoferrate(II) and hexacyanoosmate(II) showing very similar lattice constants (56). This structure also consists of a three-dimensional framework with the characteristic sequence, Ru–C–N–Mn, deviating slightly from linearity. Contrary to the Prussian blue analogs, the coordination sphere of ruthenium as well as that of manganese has a definite unique composition. Moreover, the structure is ordered, and there are no fractional occupancies of certain crystallographic positions. Ruthenium is coordinated by six carbon ends of the cyanide ions, manganese by three nitrogen atoms and three oxygen atoms of water molecules. The latter coordination octahedron is of the facial type. Two such octahedra are linked to a dinuclear unit $Mn_2N_6(H_2O)_4$ with two water molecules acting as bridging ligands.

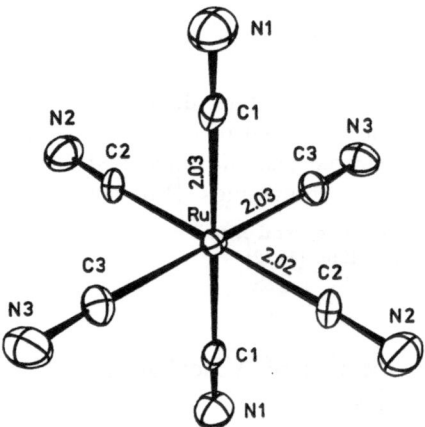

Fig. 3. A perspective drawing of the $Ru(CN)_6$ group in $Mn_2[Ru(CN)_6] \cdot 8 H_2O$

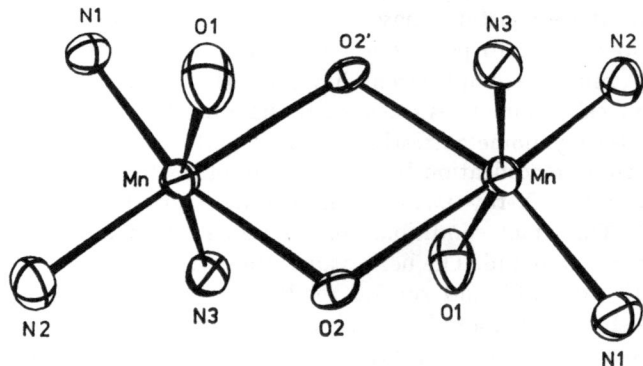

Fig. 4. A perspective drawing of the binuclear $Mn_2N_6(H_2O)_4$ group in $Mn_2[Ru(CN)_6] \cdot 8 H_2O$

The structural relationship of this monoclinic structure to the cubic one of the Prussian blue analogs is not very obvious. A comparison can be made if the dinuclear $MnN_6(H_2O)_4$ group is replaced by a hypothetical MN_6 octahedron and the monoclinic unit cell deformed to a cubic one. The resulting hypothetical structure is cubic face-centered with positions 4a and 4b alternatively occupied by Ru and M (56).

Table 5. *Lattice constants of* $Mn_2M(CN)_6 \cdot 8\,H_2O$

	$Mn_2Fe(CN)_6 \cdot 8\,H_2O$	$Mn_2Ru(CN)_6 \cdot 8\,H_2O$	$Mn_2Os(CN)_6 \cdot 8\,H_2O$
a (Å)	9.44	9.48	9.54
b (Å)	7.52	7.61	7.63
c (Å)	12.40	12.49	12.55
β (deg)	98.5	98.8	98.6
d_m (gcm^{-3})	1.78	1.90	2.25
d_c (gcm^{-3})	1.78	1.90	2.21

Tetragonal unit cells have been described for $K_2CuFe(CN)_6$ ($a = 9.85\,\text{Å}$; $c = 10.50\,\text{Å}$) (57) and for $CuPd(CN)_6$ ($a = 10.23\,\text{Å}$, $c = 11.02\,\text{Å}$) (58). In neither case has a complete resolution of the structure been carried out yet.

So far, we have confined our discussion to compounds $M_k^A[M^B(CN)_6]_l \cdot x\,H_2O$ where the metal ion M^A is octahedrally or pseudooctahedrally coordinated. The ambient linkage of the cyanide ion is a decisive structural element also in silver hexacyanometalates, $Ag_kM^B(CN)_6$, where the silver ion is linearly coordinated by two nitrogen atoms (59). A similar structural behavior is exhibited by hydrogen. N—H—N groups are found in the lattices of several cyanometal acids of the general formula $H_kM^B(CN)_6 \cdot x\,H_2O$ (29, 60, 61).

A special case is represented by the compounds $Ag_kM^B(CN)_{2k}$ and $H_kM^B(CN)_{2k}$. For stoichiometric reason, all the hydrogen and silver atoms are expected to be in N—H—N (N—Ag—N) bridges. A very close structural relationship has been established for compounds $H_3M^B(CN)_6$ and $Ag_3M^B(CN)_6$ (29, 59) (cf. Fig. 5).

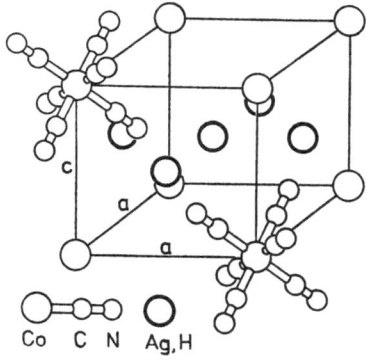

Fig. 5. The crystal structure of $Ag_3Co(CN)_6$ and $H_3Co(CN)_6$

The unit cell data of the compounds shown in Table 6 demonstrate this structural similarity between complex acids and corresponding silver salts.

Table 6. *Lattice constants of the trigonal* $H_3M^B(CN)_6$ *and* $Ag_3M^B(CN)_6$

	H		Ag	
	a	c	a	c
Cr (67)			7.12	7.48
Fe (60)	6.51	5.78		
Co (29, 59)	6.43	5.70	7.03	7.13
Rh (55)	6.56	5.89	7.01	7.35
Ir (55)	6.61	5.86		

It would be interesting to investigate hydrogen-silver pairs of other stoichiometries in order to look for a generalization of the above relationship.

Pauling and *Pauling* have related the structure of $H_3Co(CN)_6$ and $Ag_3Co(CN)_6$ (Fig. 5) to the cubic cell of Prussian blue (62). The angles CNH and CNAg are 167.3° and 157.3°, respectively. Both structures may be considered as trigonally distorted cubic structures.

The N—D—N distance in $D_3Co(CN)_6$ has been found to be 2.59 Å (29). To our knowledge this is the shortest distance known in a N—H—N bond. Unit cell data and infrared spectra of a series of other $H_kM^B(CN)_{2k}$ compounds indicate the presence of similarly short hydrogen bonds in all those crystalline acids (63, 64). There are many speculations and conjectures concerning the nature of these hydrogen bonds (62—64). Neutron diffraction studies of polycrystalline $D_3Co(CN)_6$ (29, 65) could not settle the question of whether a single-minimum or a double-minimum type potential is present. Based on the results of a model calculation, a double-minimum potential well has been proposed with a very low barrier that can easily be tunneled (66).

The ambient nature of the cyanide ion plays an important role in several crystalline cyano compounds which lie, however, outside the definition of compounds covered by this article. For a review of these structures the reader is referred to Ref. 23.

IV. Survey of Electronic Spectra

1. General Classification

As far as the electronic spectra are concerned, two groups of polynuclear transition metal cyanides are easily distinguished. The first group exhibits reflectance spectra that are unambiguously assigned as superpositions of the spectra of the constituent mononuclear species, i.e. of the chromophores M^BC_6 and $M^AN_xO_{6-x}$. No strong interaction between the two metal ions is apparent in the spectroscopic properties. The spectra of the compounds belonging to the second group, however, are dominated by a very intense band in the visible part. This absorption cannot be attributed to a mononuclear chromophore of either M^A or M^B, but is certainly due to some kind of interaction between the chromophores (8). Prominent examples of these deeply colored cyanides are compounds where M^A and M^B represent different oxidation states of the same elements, like Prussian blue. On the other hand, equally intense colors are found in polymeric cyanides, where M^A and M^B are different metals, e.g. $Cu_2Ru(CN)_6 \cdot xH_2O$ and $Fe_4[Os(CN)_6]_3 \cdot xH_2O$.

A general classification scheme for the discussion of the properties of mixed valence compounds has been introduced by *Robin* and *Day* (17). Although by no means all the deeply colored transition metal cyanides $M_k^A[B^M(CN)_m]_l \cdot xH_2O$ are mixed valence compounds in the strict sense, we will use this classification scheme for the whole group. The classification of the polynuclear cyanides is made on a purely phenomenological basis using the electronic spectra as a criterion.

According to this scheme (17) the polymeric cyanides belong either to class I or class II. The first group of cyanides mentioned above belongs to class I. The positions of the absorption bands of the polynuclear compounds may be somewhat displaced compared with the spectra of the mononuclear species. No bands, however, are observed that cannot be attributed to one of the mononuclear chromophores. The electronic interaction between M^A and M^B is assumed to be negligible, and the two valences are supposed to be firmly trapped. Accordingly, the deeply colored polynuclear cyanides may be considered as class II compounds.

2. Class I Cyanides

The hydrated hexacyanocobaltates(III), -rhodates(III), and -iridates(III) of divalent metals, and many hexacyanochromates(III), -ferrates(II) and (III), -ruthenates(II) and -osmates(II) exhibit the simple spectra according to the definition of class I. A few selected data from these compounds are presented in Tables 7 and 8. No significant band shifts are observed in the spectra of the M^BC_6 chromophore. In agreement with the structural properties described in II.3, the ligand field parameter Δ of the chromophore of M^A conforms to a mixed nitrogen-oxygen coordination.

An example of a pronounced displacement of the absorption bands is represented in $Ag_3Cr(CN)_6$. The ligand field parameter Δ of the chromophore CrC_6 in this polynuclear cyanide is reduced by 4.3 kK with respect to the mononuclear $K_3Cr(CN)_6$ (67). Similar-shifts have been observed in various salts of $Cr(NCS)_6^{3-}$ (68).

Table 7. *The position of the $^3A_{2g} \rightarrow {}^3T_{2g}$ band in various Ni(II)-hexacyano-metalates*

M^B	$\bar{\nu}$ (kK)
Cr(III)	9.8
Co(III)	9.9
Fe(II)	9.3
Ru(II)	9.6
Os(II)	9.5
Pd(IV)	10.5

Table 8. *The positions of the ligand field bands of the CrC_6 and CoC_6 chromophores in various hexacyanochromates(III) and -cobaltates(III)*

M^A	Cr(III)			Co(III)	
	$^4T_{2g}$	$^4T_{1g}$	$^4T_{2u}$	$^1T_{1g}$	$^1T_{2g}$
K	26.6	32.5	38.6	31.8	38.7
Cr(II)	26.2	33.7	38.9		
Mn(II)	28.6	33.6	38.8	30.8	39.2
Fe(II)		34.0	39.1	32.3	39.4
Co(II)	27.8	34.0	38.1	32.3	38.5
Ni(II)	27.7	33.4	37.7	32.3	40.8
Cu(II)				31.2	39.2
Zn(II)	27.0	33.3	38.3		
Cd(II)	27.0	32.8	38.0	31.2	38.5

3. Class II Cyanides

A more interesting situation is found in the deeply colored cyanides. The ions M^A and M^B in the different crystallographic sites still possess distinguishable valences, as Mössbauer and infrared spectroscopic data show (17). The visible part of the spectra of the two chromophores $M^B C_6$ and $M^A N_x O_{6-x}$ is very often not recognizable because of the high intensity of the electron-transfer band. In the ultraviolet region the electronic transitions of the constituent hexacyanometalate can still be observed. The electron-transfer bands of Prussian blue analogs have been discussed in terms of a qualitative molecular orbital scheme (8, 17, 69). A schematic diagram of the relevant one-electron levels is shown in Fig. 6.

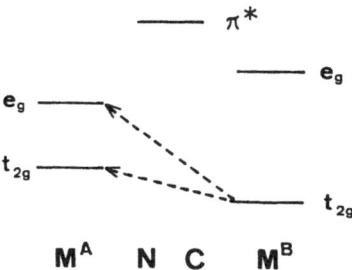

Fig. 6. A schematic diagram of the one-electron energy levels for class II cyanides

According to this approximation, the homo- or heteronuclear electron transfer corresponds to an oxidation of M^B and a reduction of M^A, e.g. $Fe^{III}-N-C-Fe^{II} \rightarrow Fe^{II}-N-C-Fe^{III}$. While this assignment seems to be obvious for the compounds where M^A is represented by Fe(III) or Cu(II), it is conceivable that the direction of the electron transfer is reversed in $Fe_3[Cr(CN)_6]_2 \cdot x H_2O$. Table 9 summarizes the position of the charge transfer bands; a few representative spectra are shown in Fig. 7.

Within the scope of the theory of mixed valence transitions (17), the extent of electron delocalization over the two different metal sites can be estimated from the intensity of the charge-transfer band. It is therefore of great importance to collect reliable data of these spectral intensities. Using the $Fe(CN)_6^{4-}$ bands as calibration, Robin found for a sol of Prussian blue an extinction coefficient of 9,800 for the 14.7 kK band (8). By measuring the transmission spectrum of a KBr pellet of $Fe_3[Cr(CN)_6]_2 \cdot x H_2O$, we obtained a value of 3,000 for the extinction coefficient ε of the band at 22 kK (70). It would be highly desirable to have suitable single crystals, which would allow a precise determination of the intensities and hence of the oscillator strengths of the electron transfer bands.

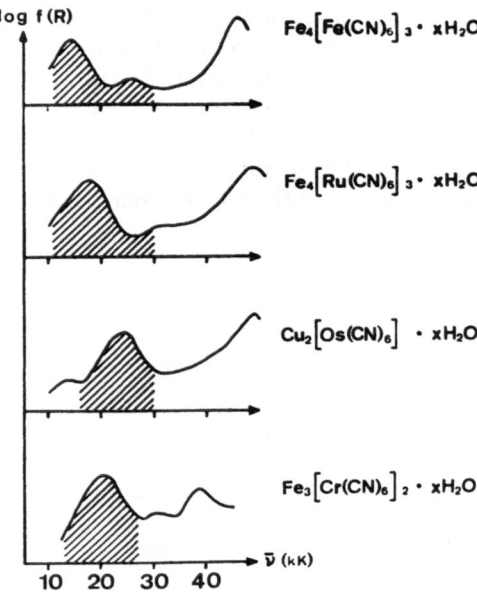

Fig. 7. Reflectance spectra of polynuclear cyanides. The shaded bands represent intervalence transitions (log f (R) = Kubelka-Munk-function)

Table 9. *The position of the electron-transfer band* $M^A \rightleftharpoons M^B$ *in various polynuclear cyanides* M_k^A $[M^B(CN)_6]_l \cdot xH_2O$

M^A	M^B	$\bar{\nu}$ (kK)
Fe(III)	Fe(II)	14.7
Fe(III)	Ru(II)	17.5
Fe(III)	Os(II)	16.8
Cu(II)	Fe(II)	21.1
Cu(II)	Ru(II)	25.0
Cu(II)	Os(II)	24.0
Fe(II)	Cr(III)	22.0
Cr(III)	Fe(II)	27.0

It would be attractive to correlate the energies of the charge-transfer bands with the electrical conductivities and the activation energies of conduction. A rather close connection of these properties has been demonstrated for $Tl_3Fe(CN)_6$ (*71*), but no obvious correlation could be observed for a series of Prussian blue analogs (*55*). It must be emphasized, however, that it is difficult to correlate spectra and conductivity for these compounds in a straightforward manner. The spectroscopic data are obtained by measuring the hydrated samples, whereas the dehydrated compounds are used to determine the conductivities. Preliminary studies show that the conductivity of the hydrated and anhydrous cyanides differ by about four orders of magnitudes.

V. Final Remarks

Despite the great number of studies carried out so far to elucidate the structure and bonding of polynuclear transition metal cyanides, our knowledge is by no means complete. One of the main reasons is certainly the fact that a great deal of experimental work has been done with rather ill-defined products. The results obtained from the investigations of these colloidal samples have therefore to be considered as not too reliable. Only in the past few years have the preparative techniques been significantly improved, permitting the growth of single crystals of several cyanides which could be used for X-ray diffraction studies. More reliable structural data began to emerge, which, amongst other things, induced a modification of the original structural model of *Keggin* and *Miles*.

While many of the crystals so far grown of polynuclear transition metal cyanides are large enough for single-crystal X-ray studies, they are still too small for neutron diffraction experiments and for spectroscopic single-crystal investigations. The electronic spectra have been determined from diffuse reflectance measurements of fine powders. Accurate extinction coefficients are not known and it has not yet been possible to take full advantage of the symmetry properties of the crystal structure for the assignment and interpretation of the spectra. In many cases the electronic spectra represent a simple superposition of the spectra of two isolated chromophores. More interesting are, of course, the compounds exhibiting a homo- or heteronuclear intervalence electron transfer, e.g. Prussian blue, copper(II)-hexacyanoferrate(II), etc. The intense electron-transfer bands of these compounds have been discussed on the basis of a one-electron molecular orbital scheme. The further improvement of the growth of single crystals is expected to stimulate precise spectroscopic studies and to lead to a more detailed picture of the electronic structure of these compounds.

Acknowledgments. The authors are indebted to the "Schweizerischer Nationalfonds" and the "Stiftung Entwicklungsfonds Seltene Metalle" for financial support.

19

IV. References

1. *Chadwick, B. M., Sharpe, A. G.:* Advan. Inorg. Chem. Radiochem. *8*, 83 (1966)·
2. *Keggin, J. F., Miles, F. D.:* Nature *137*, 577 (1936).
3. *Weiser, H. B., Milligan, W. O., Bates, J. B.:* J. Phys. Chem. *45*, 99 (1942).
4. *Ludi, A., Güdel, H. U., Hügi, R.:* Chimia (Aarau) *23*, 194 (1969).
5. *Emschwiller, G.:* Compt. Rend. *238*, 1414 (1954).
6. *Wilde, R. E., Ghosh, S. N., Marshall, B. J.:* Inorg. Chem. *9*, 2512 (1970).
7. *Davidson, D., Welo, L. A.:* J. Phys. Chem. *32*, 1191 (1928).
8. *Robin, M. B.:* Inorg. Chem. *1*, 337 (1962).
9. *Fluck, E., Kerler, W., Neuwirth, W.:* Angew. Chem. *75*, 461 (1963).
10. *Duncan, J. F., Wigley, P. W. R.:* J. Chem. Soc. *1963*, 1120.
11. *Ito, A., Suenaga, M., Ono, K.:* J. Chem. Phys. *48*, 3597 (1968).
12. *Fung, S. C., Drickamer, H. G.:* J. Chem. Phys. *51*, 4353 (1969).
13. *Erickson, N., Elliott, N.:* J. Phys. Chem. Solids *31*, 1195 (1970).
14. *Leibfritz, D., Bremser, W.:* Chemiker-Ztg. *94*, 982 (1970).
15. *Wertheim, G. K., Rosencwaig, A.:* J. Chem. Phys. *54*, 3235 (1971).
16. *Britton, D.:* Perspectives Structural Chem. *1*, 109 (1967).
17. *Robin, M. B., Day, P.:* Advan. Inorg. Chem. Radiochem. *10*, 247 (1967).
18. *Hush, N. S.:* Progr. Inorg. Chem. *8*, 391 (1967).
19. *Britton, D., Dunitz, J. D.:* Acta Cryst. *19*, 815 (1965).
20. *Rüegg, M., Ludi, A.:* Theoret. Chim. Acta *20*, 193 (1971).
21. *Rayner, J. H., Powell, H. M.:* J. Chem. Soc. *1952*, 319.
22. *Iwamoto, T., Nakano, T., Morita, M., Miyoshi, T., Miyamoto, T., Sasaki, Y.:* Inorg. Chim. Acta *2*, 313 (1968).
23. *Shriver, D. F.:* Struct. Bonding *1*, 32 (1966).
24. *Curry, N. A., Runciman, W. A.:* Acta Cryst. *12*, 674 (1959).
25. *Sequeira, A., Chidambaram, R.:* Acta Cryst. *20*, 910 (1966).
26. *Pierrot, M., Kern, R.:* Acta Cryst. B *25*, 1685 (1969).
27. *Bertinotti, C., Bertinotti, A.:* Acta Cryst. B *26*, 422 (1970).
28. *Taylor, J. C., Mueller, M. H., Hitterman, R. L.:* Acta Cryst. A *26*, 559 (1970).
29. *Güdel, H. U., Ludi, A., Fischer, P., Hälg, W.:* J. Chem. Phys. *53*, 1917 (1970).
30. *Alexander, J. J., Gray, H. B.:* J. Am. Chem. Soc. *90*, 4260 (1968).
31. *Jones, L. H., Memering, M. N., Swanson, B. I.:* J. Chem. Phys. *54*, 4666 (1971).
32. *Shriver, D. F., Shriver, S. A., Anderson, S. E.:* Inorg. Chem. *4*, 725 (1965).
33. *Brown, D. B., Shriver, D. F.:* Inorg. Chem. *8*, 37 (1969).
34. *Schneider, W.:* Einführung in die Koordinationschemie, pp. 8. Berlin–Heidelberg–New York: Springer 1968.
35. *Prout, W. E., Russell, E. R., Groh, H. J.:* J. Inorg. Nucl. Chem. *27*, 473 (1965).
36. *Brown, D. B., Shriver, D. F., Schwartz, L. H.:* Inorg. Chem. *7*, 77 (1968).
37. *Shriver, D. F., Brown, D. B.:* Inorg. Chem. *8*, 42 (1969).
38. *Ludi, A., Güdel, H. U.:* Helv. Chim. Acta *51*, 2006 (1968).
39. *van Bever, A. K.:* Rec. Trav. Chim. *57*, 1259 (1938).
40. *Rigamonti, R.:* Gazz. Chim. Ital. *68*, 803 (1938).
41. *Rollier, M. A., Arreghini, E.:* Gazz. Chim. Ital. *69*, 499 (1939).
42. *Ferrari, A., et al.:* Gazz. Chim. Ital. *89*, 2512 (1959); *90*, 1565 (1960). — Acta Cryst. *14*, 695 (1961); *15*, 90 (1962); *17*, 311 (1964).
43. *Vaciago, A., Mugnoli, A.:* Atti Accad. Nazl. Lincei Rend., Classe Sci. Fis. Mat. Nat. *25*, 531 (1958).
44. *Maer, K., Beasley, M. L., Collins, R. L., Milligan, W. O.:* J. Am. Chem. Soc. *90*, 3201 (1968).

45. International Tables for X-ray Crystallography, Vol. 1. Birmingham, England: Kynoch Press 1952.
46. *Ludi, A., Güdel, H. U., Rüegg, M.:* Inorg. Chem. *9*, 2224 (1970).
47. — *Ron, G.:* Chimia (Aarau) *25*, 333 (1971).
48. *Güdel, H. U.:* Acta Chem. Scand., in press.
49. *Ludi, A., Ron, G.:* to be published.
50. *Wolberg, A.:* Acta Cryst. B *25*, 161 (1969).
51. *Güdel, H. U., Stucki, H., Ludi, A.:* to be published.
52. *Hamilton, W. C., Ibers, J. A.:* Hydrogen Bonding in Solids. New York: W. A. Benjamin, Inc. 1968.
53. *Wyckoff, R. W. G.:* Crystal Structures, 2nd ed., Vol. III, pp. 344, 377, 382. New York: Interscience 1965.
54. *Wells, A. F.:* Structural Inorganic Chemistry, 3rd ed., p 495. Oxford: Clarendon Press 1962.
55. *Ludi, A.:* unpublished results.
56. *Rüegg, M., Ludi, A., Rieder, K.:* Inorg. Chem. *10*, 1773 (1971).
57. *Gellings, P. J.:* Z. Physik Chem. (Neue Folge) *54*, 296 (1967).
58. *Ron, G., Ludi, A.:* to be published.
59. *Ludi, A., Güdel, H. U.:* Helv. Chim. Acta *51*, 1762 (1968).
60. *Haser, R., de Broin, C. E., Pierrot, M.:* Compt. Rend. 272, 1308 (1971).
61. *Basson, S. S., Bok, L. D. C., Leipoldt, J. G.:* Acta Cryst. B *26*, 1209 (1970).
62. *Pauling, L., Pauling, P.:* Proc. Natl. Acad. Sci. U.S. *60*, 362 (1968).
63. *Evans, D. F., Jones, D., Wilkinson, G.:* J. Chem. Soc. *1964*, 3164.
64. *Beck, W., Smedal, H. S.:* Z. Naturforsch. *20b*, 109 (1965).
65. *Güdel, H. U., Ludi, A., Fischer, P.:* J. Chem. Phys. *56*, 674 (1972).
66. — J. Chem. Phys., in press.
67. *Kirk, A. D., Schläfer, H. L., Ludi, A.:* Can. J. Chem. *48*, 1065 (1970).
68. *Wasson, J. R., Trapp, C.:* J. Inorg. Nucl. Chem. *30*, 2437 (1968).
69. *Braterman, P. S.:* J. Chem. Soc. *1966A*, 1471.
70. *Stucki, H., Ludi, A.:* unpublished results.
71. *Braterman, P. S., Phipps, P. B. P., Williams, R. J. P.:* J. Chem. Soc. *1965*, 6164.

Received April 13, 1972

Electronic Spectra of Tetrahedral Oxo, Thio and Seleno Complexes Formed by Elements of the Beginning of the Transition Groups

A. Müller and E. Diemann

Institut für Chemie der Universität, D-46 Dortmund

C. K. Jørgensen

Département de Chimie physique, Université, CH–1211 Genève

Table of Contents

1. Comparison with Halide Complexes

The simultaneous presence of a reducing ligand and an oxidizing central atom in a complex produces *electron transfer bands* in the visible or the near ultra-violet. Such transitions were recognized by *Rabinowitch* (1) in halide complexes of iron(III) $Fe(H_2O)_5X^{+2}$ and by *Linhard* and *Weigel* (2) in cobalt(III) pentammines $Co(NH_3)_5X^{+2}$. When $X = Cl$, Br and I, a regular decrease toward lower wave-numbers of the electron transfer bands is observed when the halide becomes more reducing, whereas $X = F$ does not produce a specific electron transfer band because the water or ammonia molecules already are more reducing than fluoride; and $Fe(H_2O)_6^{3+}$ has an electron transfer band at 42 kK (1 kilo-kayser = 1000 cm⁻¹) and $Co(NH_3)_6^{+3}$ at 50 kK.

These qualitative observations go, in part, back to *Fromherz* (3) but they were connected with the "ligand field" treatment of internal transition of the partly filled d-shell and with the group-theoretical conditions for the symmetry types of the energy levels in a study (4—6) of the hexahalide complexes MX_6^{+z-6} of the central atom M in the oxidation state $+z$. It was possible to introduce the *optical electronegativity* x_{opt} by the definition

$$\nu_{corr} = [x_{opt}(X) - x_{opt}(M)] \cdot 30 \text{ kK} \tag{1}$$

and an uncorrected optical electronegativity (of frequent use in the treatment of $4f$-group complexes (5, 7))

$$\nu_{uncorr} = [x_{opt}(X) - x_{uncorr}(M)] \cdot 30 \text{ kK} \tag{2}$$

In Eq. (2), the wave-number ν_{uncorr} of the first intense (*Laporte*-allowed) electron transfer band is directly inserted, whereas ν_{corr} in Eq. (1) has been corrected for spin-pairing energy (and in the case of f-group complexes, also for other effects of interelectronic repulsion and for the first-order relativistic effect usually called "spin-orbit coupling"). These relatively minor corrections (usually below 8 kK) are not of great importance for the present review because they vanish for central atoms having no d-electrons in the groundstate, and hence, $x_{opt}(M)$ and $x_{uncorr}(M)$ coincide.

The choice 30 kK of the numerical coefficient in Eqs. (1) and (2) allows the optical electronegativities x_{opt} of the four halide ligands

$$F^- \ 3.9 \qquad Cl^- \ 3.0 \qquad Br^- \ 2.8 \qquad I^- \ 2.5 \tag{3}$$

to have the same values as *Pauling*'s electronegativities. Different types of electronegativities have been theoretically discussed (8). The result Eq. (3) is by no means trivial, showing among other consequences that for a highly different choice of central atoms, the difference of reducing character between two halide ligands remains roughly constant. Once the values of x_{opt} have been established for a large number of central atoms M^{+z} (in the d-groups, x_{opt} usually increase about 0.2 unit for each increase of one unit of the oxidation state of a given element, though the increase from M(II) to M(III) is larger than from M(V) to M(VI)) it is possible to evaluate the optical electronegativities of other ligands than the four halides. Thus, $x_{opt} = 3.5$ for H_2O, 3.3 for NH_3, 3.2 for SO_4^{-2} (9), 2.8 for N_3^- (10) and 2.7 for acetylacetonate bound by two oxygen atoms to the central atom (11).

Oxygen-containing ligands show a considerable spreading, 0.8 unit from water to acetylacetonate. This variation is not uniquely determined by the chemically reducing character of the ligand; as first pointed out by *Barnes* and *Day* (9) hypophosphite $H_2PO_2^-$ has $x_{opt} = 3.3$, slightly higher than for sulphate, whereas, obviously, hypophosphite is chemically far more reducing. The reason for this discrepancy is that the optical transitions corresponding to the electron transfer spectra obey *Franck* and *Condon's* principle, the internuclear distances do not have the time to re-arrange in less than 10^{-13} sec, whereas the reducing character of $H_2PO_2^-$ is connected with the two hydrido ligands on the rear side of the phosphorus(V) atom having its two (or one of the) oxo ligands bound directly to the central atom accepting the electron in the excited state. Besides the fact that x_{opt} of H^- is not particularly low and seems to have the order of magnitude 3.2 in the rhenium(VII) complex ReH_9^{-2} (*6, 12*) the electron transfer from a long distance (say from the hydride part of $EuO_2PH_2^{+2}$ to the oxidizing central atom europium(III)) is discouraged by the kind of interelectronic repulsion effect called *charge separation effects* (*8*). They contribute most definitely to the *anisotropic behaviour* of uranyl complexes (*13*) where $UO_2X_n^{+2-n}$ consists of two oxo ligands bound very strongly at a short distance perpendicular on the equatorial plane of the complex containing the uranium(VI) central atom connected with long distances to the n (usually 4, 5 or 6) weakly bound ligands X^-. Thus, the equatorial ligands of uranyl complexes indicate $x_{uncorr} = 1.8$ of the central atom in sharp contrast to the value 2.4 obtained from octahedral UCl_6 or by extrapolation from 2.6 for NpF_6 and 2.85 for Pu(VI). Said in other words, the charge separation effects have decreased the apparent reducing character of the equatorial ligands X^- to the extent of 0.6 unit, more than the usual difference between chloride and iodide.

When evaluating the optical electronegativity of *oxide*, it is a pity that the anisotropic character of the uranyl ions prevent a direct estimate from the (weak $\pi_u \rightarrow 5f$) electron transfer band at 22 kK producing the luminescence of uranyl ions in aqueous solution or in salts containing bidentate nitrate groups in the equatorial plane such as $UO_2(O_2NO)_3^-$. The purple colours produced by traces of praseodymium(IV) or terbium (IV) in thorium (IV) oxide (crystallizing in fluorite lattice) can be described (*14*) by $x_{opt} = 3.1$ for O^{-2} but this value might conceivably depend on the strength of the *Madelung* potential (the surrounding cations stabilize this ligand which would loose an electron spontaneously in the gaseous state) and also on the question whether all the lone-pairs of oxide are occupied by forming σ-bonds (*15, 16*). It was seen above that azide having two lone-pairs has x_{opt} 0.5 unit lower than ammonia containing only one lone-pair used for σ-bonding to the central atom. As a matter of fact, the σ-orbital formed by the loosest bound *p*-shell of the halide ligand has 12 to 15 kK higher excitation energy in electron transfer spectra (*4, 6*) than the two π-orbitals, corresponding to x_{opt} for the σ-orbital effectively, being 0.4 to 0.5 unit higher than the values given in Eq. (3).

25

The electron transfer bands of oxo complexes should not be confused with the $2p^6 \rightarrow 2p^5\,3s$ transitions producing the orange colour of caesium(I) oxide Cs_2O or a band in the near ultra-violet, at 32.8 kK, of BaO (17), whereas it has moved to 45 kK in the isoelectronic La_2O_3 and to even higher wave-numbers in Al_2O_3 and BeO. Hence, this transition is highly influenced by the *Madelung* potential in contrast to the similar $5p^6 \rightarrow 5p^5\,6s$ transitions of iodides (5).

Most solid transition-group sulphides are either metallic or dark coloured. The corresponding strong absorption in the visible may be due either to electron transfer between atoms of the same element in differing oxidation states in non-stoichiometric compounds or due to collective effects like the crystalline salts of $Pt(CN)_4^{-2}$ (18). Sulphur-containing ligands such as the bidentate $(C_2H_5)_2NCS_2^-$ and $(C_2H_5O)_2PS_2^-$ have x_{opt} values scattered around 2.7 (19, 20) but the only clear-cut case of a solid with sulphur-bridges is lemon-yellow ytterbium(III) sulphide for which 2.5 is estimated (21). To the chemist, it is obvious that sulphur-containing ligands under equal circumstances are more reducing than oxygen-containing ligands. A numerical difference of 0.6 unit is quite reasonable, compared with 0.9 unit between F^- and Cl^-. It is also well-known that F and Cl are more different, electronegativity-wise, than O and S.

This attenuation of the electronegativity variation in the earlier columns of the Periodic Table is even more pronounced when S is compared with Se. A choice of selenium-containing ligands (20, 22, 23) have x_{opt} 0.05 to 0.1 unit below that of the corresponding sulphur-containing ligands. Before the work described in this review, the spectra of complexes of Se^{-2} were not known.

2. Molecular Orbitals in Tetrahedral Complexes

Historically speaking, permanganate MnO_4^- was the earliest tetrahedral transition-group complex to be treated by M.O. theory, and a very large number of attempts of numerical calculations on this and related species have followed. Since the shell which contains one electron in green MnO_4^{-2} and two electrons in blue MnO_4^{-3} is empty in MnO_4^-, the purple colour is due entirely to electron transfer bands. This is a considerable simplification because cobalt(II) tetrahalides CoX_4^{-2} are known to have internal $3d^7$ transitions not much weaker than allowed electron transfer bands, and actually, the blue colour of MnO_4^{-3} is due (24) to internal $3d^2$ transitions. Most chemists believe that the intensity of the absorption bands in the visible of permanganate is unusually large, but actually, the *oscillator strength* P is only 0.03 corresponding to a moderately strong electron transfer band. As seen from a compilation of P-values *(25)* really strong bands of hexahalide complexes are due to transitions from (mainly σ) t_{1u} orbitals to the upper, σ-anti-bonding sub-shell e_g and have P between 1 and 1.5. However, these bands occur in the ultra-violet above 39 kK.

Wolfsberg and *Helmholz (26)* applied their model (for a critical discussion, see refs. *(8)* and *(27)*) to CrO_4^{-2} and MnO_4^-. Surprisingly enough, the sub-shell energy difference Δ between the sub-shell of symmetry type e in the point-group T_d and the sub-shell of symmetry type t_2 turned out to be positive, like in octahedral chromophores MX_6, and not like the well-known tetra-halides of cobalt(II) and many other 3d-group central atoms. In the orientation of the regular tetrahedron where the equivalence of the three Cartesian axes is most readily seen (the four ligand nuclei occupying every second of the eight corners of a cube) the former sub-shell consists of the two orbitals having angular functions proportional to $(x^2 - y^2)$ and $(3z^2 - r^2)$ and the latter sub-shell consists of the three orbitals having angular functions proportional to (xy), (xz) and (yz). However, it was realized around 1959 from a variety of physical measurements reviewed by *Carrington* and *Symons (28)* that Δ is indeed negative in the d-group tetroxo complexes, and that the lower sub-shell contains one electron in each of the two orbitals in MnO_4^{-3} producing the groundstate $^3A_2$ which is stable toward *Jahn-Teller* distortions. Usually, it is not possible to make tetroxo complexes containing more than two d-like electrons, with the exception of the cobalt(V) complex CoO_4^{-3} having a magnetic moment *(29)* corresponding to $S = 2$ and presumably containing two electrons in the lower sub-shell (e) and two electrons in the upper sub-shell (t_2). Incidentally, the fact that this complex is not diamagnetic ($S = 0$) with all four electrons in the lower sub-shell shows that $-\Delta$ is not larger than half the spin-pairing energy or $3D$ *(7)* presumably about 10 kK by comparison with MnO_4^{-3} *(24)*.

Though it is firmly established that Δ is negative, the numerical value has been subject for intensive discussion. Two extremes are represented by *Gray (30)* believing Δ is -25 kK in MnO_4^- and *De Michelis et al.* *(31)* deriving an almost vanishing value from the electron transfer spectrum of permanganate, introducing a considerable amount of M.O. configuration intermixing in the excited levels $^1T_2$. One of the writers *(6)* argues that a moderate value -10 kK is appropriate (perhaps modified from a value between -12 and -15 kK by charge separation effects). It is beyond discussion *(32)* that Δ is more negative than -20 kK in the $4d^2$ system RuO_4^{-2} but it is generally known *(7)* that the sub-shell energy differences are some 50 percent larger in 4d-group complexes than in the corresponding 3d-group complexes.

There is general agreement today that the order of M.O. energies in tetroxo complexes is

$$(d)t_2 > (d)e > (\pi)t_1 > (\pi)t_2 > [(\pi)e] > (\sigma)t_2 > [(\sigma)a_1] \qquad (4)$$

where the main constituent (central atom d; oxygen $2p\pi$; oxygen σ) is indicated in parenthesis. The sharp brackets indicate orbitals from which transitions have not been identified with certainty. The reasons are that $(\pi)e \to (d)e$ and $(\sigma)a_1 \to (d)e$ do not have electric dipole moments and hence are symmetry-forbidden transitions in the point-group T_d and that $(\sigma)a_1 \to (d)t_2$ corresponds to a very high wave-number.

The first transitions correspond to $(\pi)t_1 \to (d)e$ in d^0 systems containing no $(d)e$ electrons in the groundstate. The excited M.O. configuration involving five $(\pi)t_1$ and one $(d)e$ electron consists of four terms, $^3T_1$, $^3T_2$, $^1T_1$ and $^1T_2$ containing, all together, $6 \times 4 = 24$ states. As we shall see below, there is very little evidence for a perceptible energy separation between these four terms of the same M.O. configuration in most of our complexes of four chalkogen ligands. However, before the main maximum of MnO_4^- at 18.6 kK (due to the symmetry-allowed $^1A_1 \to {}^1T_2$) a shoulder occurs at 16 kK, and electronic origins can be detected at 14.37, 14.44 and 14.48 kK (33). All of these bands show a pronounced vibrational structure, which is also known from gaseous RuO_4 (34) and OsO_4 (35) and from TcO_4^- and ReO_4^- in aqueous solution (36) whereas most of the other complexes do not show a vibrational structure. The relatively weak permanganate bands in the red (33) do not change their intensity to any large extent going from 300 °K to 4 °K. This observation does not suggest that the excited level $^1T_1$ has vibronic coupling with the symmetry-allowed transition to $^1T_2$ in which case the intensity should increase with increasing temperature. There are two essentially different explanations possible; either the spin-forbidden transitions to $^3T_1$ and $^3T_2$ are exceedingly weak (but *Jørgensen* has not been able to detect them below 14 kK in the reflection spectrum of undiluted $KMnO_4$) and the bands in the red are due to $^1T_1$ obtaining their intensities from a systematic deviation from regular tetrahedral symmetry T_d on an instantaneous picture; or otherwise, as considered more probable by the writers, the bands in the red are one or both of the spin-forbidden transitions. The wave-number spreading 4 kK between the four terms of the M.O. configuration is smaller than of the corresponding levels belonging essentially to $(t_{2g})^5(e_g)$ of $Co(NH_3)_6^{+3}$, viz. $^3T_{1g}$ at 13.0 kK, $^3T_{2g}$ at 17.2 kK, $^1T_{1g}$ at 21.0 and $^1T_{2g}$ at 29.5 kK. However, the parameters of interelectronic repulsion producing the spreading 16.5 kK in the excited configuration of the cobalt(III) complex belong to two 3d-like sub-shells and are expected to be far larger than in the case of electron transfer spectra. Thus, the osmium(IV) hexahalide complexes OsX_6^{-2} have almost the same shape of the detailed structure as the corresponding OsX_6^{-3} and IrX_6^{-2} (4, 37) though the two latter, isoelectronic, cases have no splittings due to interelectronic repulsion. We have found no evidence for precursors to the first strong band of MoO_4^{-2} and TcO_4^- by plotting log ε as a function of the wave-number (this gives a parabola when the molar extinction coefficient ε varies as a Gaussian error-curve). It may be mentioned that the $4d^6$ complex $Rh(NH_3)_6^{+3}$ has $^3T_{1g}$ at 27 kK, $^1T_{1g}$ at 32.7 and $^1T_{2g}$ at 39.1 kK, showing somewhat smaller effects of interelectronic repulsion.

In the $3d^0$ tetroxo complexes, the two next transitions $(\pi)\,t_1 \to (d)\,t_2$ and $(\pi)\,t_2 \to (d)\,e$ almost coincide. One of the writers (25) suggested that the broad shoulder at 28.5 kK without vibrational structure, of MnO_4^-, is due to the former transition to the σ-anti-bonding upper sub-shell, whereas the band centered at 32.3 kK with pronounced vibrational structure is the latter transition, suggesting a difference of excitation energy 14 kK between $(\pi)\,t_1$ and $(\pi)\,t_2$ of Eq. (4). The presence of two distinct electronic transitions have been confirmed in absorption spectra of crystals at 4 °K (38). On the other hand, the difference of excitation energy between $(\pi)\,t_1$ and $(\pi)\,t_2$ is definitely (32) only 6 kK in the case of TcO_4^- and RuO_4.

The order (4) has been confirmed (39) for CrO_4^{-2} and MnO_4^- using the *Faraday effect* (40), i.e. the circular dichroism obtained from a strong external magnetic field. It is by no means trivial that $(\pi)\,t_1$ is easier to excite than $(\pi)\,t_2$ because the symmetry type t_2 is common for one set of three π orbitals and one set of three σ orbitals, and M.O. theory would allow them to mix *via* non-diagonal elements of the effective one-electron operator between orbitals centered on differing ligands. As a matter of fact, this ligand-ligand repulsion (41) is sufficiently strong in hexahalides such as OsX_6^{-2} and IrX_6^{-2} to make the highest t_{1u} (having about three-quarters π and one-quarter σ character) more readily excited than the pure π orbitals t_{2u}. This has been demonstrated by the *Faraday* effect on solutions (42) and crystals (43) and by the recent parametrization of the energy levels (44). Apparently, the fact that the X—X distances in regular tetrahedral MX_4 are $2/3\sqrt{6} = 1.63$ times the M—X distance whereas the shortest X—X distances in regular octahedral MX_6 are only $\sqrt{2} = 1.41$ times the M—X distance produce much larger ligand-ligand anti-bonding effects in hexahalides than in tetroxides and tetrahalides. It is remembered (41) that such anti-bonding tends to increase about 7 percent when the internuclear distance is 1 percent shorter.

The tetrahedral molecules $TiCl_4$, $TiBr_4$ and TiI_4 are also $3d^0$ systems, and measurements of the *Faraday* effect (45) show a close analogy with MnO_4^- and are incompatible with an opinion expressed by some theorists that the $4s$-like totally symmetric (a_1) orbital has lower energy than the two $3d$-sub-shells. The fractional charge of the titanium(IV) central atom in $TiCl_4$ is about $+2$ (27). $TiCl_4$ has the same difficulty as VO_4^{-3} and CrO_4^{-2} that the second absorption band seems to be due to the almost coinciding transitions $(\pi)t_1 \to (d)t_2$ and $(\pi)t_2 \to (d)e$. It is interesting to compare the electron transfer bands (wave-numbers in kK) of the tetrahedral molecules (46) with the first symmetry-allowed electron transfer band of octahedral $TiCl_6^{-2}$ and $TiBr_6^{-2}$ (47) and TiI_6^{-2} (48):

$$
\begin{array}{llll}
TiCl_4 & 35.8 & TiCl_6^{-2} & 29.7 \\
TiBr_4 & (28.0),\ 29.5 & TiBr_6^{-2} & 21.8 \\
TiI_4 & 19.6 & TiI_6^{-2} & 12.1,\ 14.3
\end{array}
\tag{5}
$$

It is seen that octahedral Ti(IV) seems somewhat more oxidizing, $x_{opt} = 2.0$ to 2.1, than tetrahedral Ti(IV) having $x_{opt} = 1.85$. This difference may, in part, be ascribed to the stronger ligand-ligand anti-bonding effects in the hexahalides, but this can only express a part of the truth, because the difference in excitation energy between $(\pi)t_{1g}$ and $(\pi)t_{2u}$ which is a good measure of ligand-ligand repulsion in hexahalides (49) is known to be 10 kK for hexafluorides and 7 kK (0.23 electronegativity unit) for hexachlorides, and $(\pi + \sigma)t_{1u}$ show a smaller effect whereas, on the other hand, $(\pi)t_1$ must be somewhat ligand-ligand anti-bonding in tetrahedral complexes.

It cannot be argued that the differences in Eq. (5) are due to an intrinsic tendency toward lower wave-numbers for a given combination of central atom and ligand when the coordination number N increases. Thus, N is undoubtedly larger than six (probably 8 or 9) in the complexes $SmBr^{+2}$ (40.2 kK), $EuBr^{+2}$ (31.2 kK), $TmBr^{+2}$ (44.5 kK) and $YbBr^{+2}$ (35.5 kK) in almost anhydrous ethanol (50) but the wave-number (51) of the first electron transfer band is $SmBr_6^{-3}$ (35.0), $EuBr_6^{-3}$ (24.5), $TmBr_6^{-3}$ (~ 38.6) and $YbBr_6^{-3}$ (29.2). This difference around 6 kK in the opposite direction of Eq. (5) can be explained along two lines of thought; either there is no bromide-bromide repulsion in the former series, or the empty $3d$ orbitals of tetrahedral titanium(IV) are all five strongly destabilized. It may be noted that anhydrous $EuBr_3$ (having $N = 9$) is distinctly less yellow than the octahedral complex. A further argument for the second alternative is that certain organometallic compounds containing titanium-carbon bonds are colourless. This may be compared with the strong yellow colour of $PbCl_4$ and $PbCl_6^{-2}$ where the electron transfer takes place to the empty σ-anti-bonding $6s$ orbital (4) whereas $Pb(CH_3)_2Cl_2$ and even $Pb(CH_3)_2Br_2$ are colourless, the empty, totally symmetric orbital presumably being more anti-bonding.

When optical electronegativities are evaluated for M(II) and M(III) tetrahalides (5, 52) roughly the same variation is found as one would obtain by extrapolation from the hexahalides. However, a certain analogy to Eq. (5) can be seen from the fact that $x_{uncorr} = 2.1$ for Fe(III) in $FeCl_4^-$ and, after correction for spin-pairing energy, $x_{opt} = 2.5$ to be compared with 2.4 for the lower (filled) sub-shell of octahedral cobalt(III). Unfortunately, it is not possible to derive an exact value of x_{opt} from the reflection spectra of the orange salts of $FeCl_6^{-3}$ only known in solids, but it would be surprising if it was below 2.4. The electron transfer spectra of $3d$-group tetrahalides (53, 54) show narrow and not particularly intense bands. It has only been possible to identify the transitions from $(\sigma)t_2$ to $(d)t_2$ as some shoulders in NiX_4^{-2} (6) and a (not very strong) band in $CuCl_4^{-2}$ and $CuBr_4^{-2}$ (in crystals, the symmetry is D_{2d} rather than T_d which contributes to the lower intensity of a transition to the half-filled $(x^2 - y^2)$ orbital) because even ZnX_4^{-2} show extremely strong bands of another origin (55). This may either be electron transfer to the $(s) a_1$ orbital like in tetrahedral mercury(II), thallium (III) and tin(IV) complexes or intra-halogen excitations of the type $4p^6 \rightarrow 4p^5 5s$ in bromide and $5p^6 \rightarrow 5p^6 6s$ in iodide (5). There is no sign of comparable $2p^6 \rightarrow 2p^5 3s$ excitations of oxide in tetroxo complexes. On the other hand, the late Dr. R. C. Hirt (36) observed a strong band of MnO_4^- at 52.9 kK ($\varepsilon_{max} = 20200$) which most probably is the $(\sigma)t_2 \rightarrow (d)t_2$ excitation in spite of the oscillator strength P being only 0.6, about half the value for the similar transition in hexahalides (25).

3. Mixed Complexes with Three Oxide and a Fourth Ligand

In the case of hexahalides, the most carefully studied mixed complexes are all the ten geometrical isomers of $OsCl_nBr_{6-n}^{-2}$ and $IrCl_nBr_{6-n}^{-2}$ (49) and of $OsCl_nI_{6-n}^{-2}$ (56). The theoretical treatment (49, 57) is complicated by the fact that the orbital energy differences are influenced by effects of ligand-ligand repulsion to an extent which is not much smaller than the intrinsic differences of optical electronegativity.

In tetrahedral complexes, the problems related to ligand-ligand repulsion seem less serious, and one might have expected that mixed complexes of a halide and of oxide would allow the estimate of x_{opt} of the oxide. Actually, the modification of the spectrum of yellow CrO_4^{-2} by protonation or by substitution of other ligands to form orange $CrO_3(OH)^-$, CrO_3F^- and CrO_3Cl^- was already, discussed by *Helmholz, Brennan* and *Wolfsberg* (58). It is obvious that the similarity between the three latter spectra cannot be explained by F^- being more reducing than O^{-2}. The idea that x_{opt} of oxide is increased to about 3.9 by the strong *Madelung* potential might have been supported by the fact that gaseous OsF$_6$ has the first *Laporte*-allowed electron transfer band at 35.7 kK corresponding to $x_{uncorr} = 2.7$ and $x_{opt} = 2.6$ for Os(VI) whereas OsO$_4$ has the baricenter of the first band at 33.8 kK in the gaseous state and at 34.6 kK in aqueous solution. One can extrapolate from the regular behaviour of the hexahalides that x_{opt} for Os(VIII) would be between 2.7 and 2.8 corresponding to an approximate value of 3.9 for oxide. However, this argument is defectious because chloride substitution in a tetroxo complex should shift the first electron transfer band 27 kK toward lower wavenumbers what, most definitely, is not the case.

The solution to this problem was suggested by *Carrington* and *Jørgensen* (32). The oxide ligand must have an extraordinarily large *π-anti-bonding effect* on the central atom d-shell, and specifically, larger than of OH$^-$ and of the halides. Other cases of strong π-anti-bonding effect of one oxide ligand are known from the vanadyl ion VO(H$_2$O)$_4^{+2}$ and MoOX$_5^{-2}$ (59) or of two oxide ligands in many rhenium(V) complexes such as ReO$_2$(NH$_3$)$_4^+$ (60). Interestingly enough, the separation between the non-bonding (xy) and the π-anti-bonding (xz, yz) orbitals in all these complexes is between 12 and 15 kK or between 0.4 and 0.5 units of optical electronegativity. It is not easy to know what x_{opt} of Cr(VI) "ought to be", an extrapolation from Ti(IV), Zr(IV) and Mo(VI) suggests a value close to 2.6 compatible with the yellow colour of CrF$_6$ prepared at low temperature by *Glemser et al.* (61). The shift of the first symmetry-allowed transition from 26.8 kK in CrO$_4^{-2}$ to 22.2 kK in CrO$_3$F$^-$ is compatible with $x_{opt} = 3.0$ for oxide and a modification (due to the π-anti-bonding effect) of x_{opt} of Cr(VI) to 2.1 when four oxide ligands are present, and to 2.25 when only three oxide ligands are available. Actually, the three values might also be 2.9, 2.0 and 2.15 because the purple CrO$_2$Br$_2$ studied by *Krauss* and *Stark* (62) has an electron transfer band at 20.0 kK. This would be consistent with the optical electronegativities 2.8 for the bromide ligands and 2.15 for the central atom modified only by the π-anti-bonding effect of the two oxide ligands. Recently, salts of CrO$_3$Br$^-$ have been prepared (63) showing a weaker band in acetone solution at 22.3 kK and a stronger band with vibrational structure at 27.6 kK. In acetonitrile, these bands occur at 22.4 and 27.9 kK, and further on, a shoulder at 34.3 kK is observed on a stronger band (not showing vibrational structure) at 38.4 kK. The two strong bands are not essentially different from 28.9 and 36.4 kK for CrO$_3$F$^-$ and 28.1 and 35.1 kK for CrO$_3$Cl$^-$ in CH$_3$CN. However, the non-monotonic evolution from 36.4 and 35.1 to 38.4 kK in CrO$_3$Br$^-$ makes it probable that the latter band is due to a M.O. consisting mainly of bromine $4p\sigma$. The analogous excitation of chlorine $3p\sigma$ in CrO$_3$Cl$^-$ may correspond to a band observed at 41.4 kK in acetonitrile solution since the three strongest bands of CrO_4^{-2} in aqueous solution occur at 26.8, 36.6 and 48.8 kK.

The bridging oxygen atom in dichromate $O_3CrOCrO_3^{-2}$, which is thermodynamically much more stable in aqueous solution than the analogous pyrosulphate, has almost the same effect as protonated OH^- and a similar weaker π-anti-bonding effect. *Butowiez (64)* succeeded in identifying the absolutely lowest electronic excited level in crystalline dichromates at 4 °K as a triplet close to 18.2 kK; this observation might suggest that the "precursor band" of CrO_4^{-2} at 24 kK analogous to the 16 kK absorption of MnO_4^- is a spin-forbidden transition to a triplet. Other authors *(33)* insist that the electronic origins close to 14.5 kK of MnO_4^- in $KClO_4$ must be due to $^1T_1$ but the absence of absorption lines at lower wave-numbers due to $^3T_1$ and $^3T_2$ is surprising, and the polarization measurements do not exclude some Γ_J components of the triplets.

"Permanganic acid" is a mixed trioxo-monohydroxo complex $MnO_3(OH)$ which can be seen as a green colour when MnO_4^- is dissolved in sufficiently strong perchloric acid. It has an absorption spectrum closely similar to that of gaseous MnO_3F *(65)* having a weaker band at 15.0 kK and a stronger band with vibrational structure at 22.4 kK in close analogy to CrO_3Cl^- having the weak band at 22.75 and the structured band at 28.4 kK. In both cases *(32)* the weak band can be ascribed to the transition from the a_2 component (in the point-group C_{3v}) originating from $(\pi)t_1$ (in T_d) to the lower empty sub-shell of the central atom whereas the second band corresponds to the transition from the two degenerate e-orbitals (in C_{3v}) originating from $(\pi)t_1$ (in T_d) to the same, lower sub-shell (remaining degenerate in C_{3v}). It may be noted that the baricenter of these transitions (weighted by the factors 1 and 2) occurs slightly above $^1T_2$ in MnO_4^- and almost coincides in CrO_4^{-2}. *Briggs (66)* studied MnO_3Cl and found excited levels at 16.0, 20.8 and 30.5 kK. It is remarked that all of these transitions fall 7 to 8 kK lower in the manganese(VII) than in the corresponding chromium(VI) complex. Hence, if optical electronegativities can be defined at all for these central atoms, they are 0.25 unit lower for Cr(VI) than for Mn(VII). *Royer (67)* suggested that the green solution of MnO_4^- in concentrated sulphuric acid contains planar MnO_3^+. We believe that these species contain tetrahedrally coordinated Mn(VII), for instance as $O_3MnOSO_3^-$ or protonated forms. It is true that Cr(VI) readily form half-esters O_3CrOR^- with alcohols ROH *(68)* and that solutions in sulphuric and phosphoric acid have slightly differing spectra suggesting the existence of mixed anions. However, all these spectra are sufficiently similar to indicate O_3MX of the local symmetry C_{3v}.

The a_2 component of the $(\pi)t_1$ orbitals in MO_3X is concentrated equally on the three oxide ligands and cannot involve π orbitals of the X ligand. It has three oxygen-oxygen node-planes, and its lower excitation energy compared with the unsubstituted tetroxo complex must be ascribed to higher energy of the empty sub-shell in the latter case. On the other hand, the two e-orbitals contain each a proportion of one of the two π orbitals of X. Supposing that X has a very high electronegativity, such as fluoride, the excitation energy of the e component is increased for this reason. However, in the case of X = Br or I, one would expect the e-component to have lower excitation energy because of this $X\pi$ character, though the distance between a_2 and e is still 5.5 kK in CrO_3Br^- and 4.8 k K in MnO_3Cl. The corresponding values are 6.7 kK in CrO_3F^- and 7.4 kK in MnO_3F. It might be argued that repulsion between the oxo ligands (41) contributes to the low excitation energy of a_2.

A rather striking case of a substituent is the nitrido-osmate(VIII) OsO_3N^- (69). One of the writers (70) previously demonstrated that not only the oxide ligands (this is well-known from OsO_4) but also the nitride ligand has lost its proton affinity in aqueous solution, and that the lemon-yellow OsO_3N^- has the same spectrum in 1 molar perchloric acid as in aqueous ammonia. Said in other words, the imido complex $OsO_3(NH)$ is a very strong acid. The vibrational components at 27.5 and 28.3 kK are moderately weak ($\varepsilon \sim 200$) and might either represent a precursor band, a transition from the a_2 or from the e component of $(\pi)t_1$. The interesting point is that the shift toward lower wave-numbers is not more than 4 kK relative to a more intense group with vibrational structure centered at 31.5 kK ($\varepsilon \sim 1000$) followed by a shoulder at 35.8 kK and a broad band ($\varepsilon \sim 2500$) at 44 kK. It is remarkable to what extent these bands are weaker than those of OsO_4 by a factor two to four. The oscillator strengths of transitions in CrO_3X^- are also weaker than of CrO_4^{-2}; there is no obvious theoretical explanation of this fact. The conclusions one may draw from the band positions of OsO_3N^- is that nitride is, at most, 0.2 electronegativity unit more reducing than oxide in this specific case. Actually, the most plausible assignment of the band at 31.5 kK to the e component of $(\pi)t_1$ corresponds to a much closer similarity between nitride and oxide ligands, unless N^{-3} produces an even stronger π-antibonding effect on the lower $5d$ sub-shell of the central atom than O^{-2}. If the electronegativity difference between oxygen and nitrogen is important, one expects the e component to be easier to excite than a_2 like in the case of MO_3S discussed below, and hence, the structure at 28 kK is not due to a_2.

4. Spectra of Complexes with Four Chalkogenide Ligands

A comparative study of the absorption spectra of tetroxo complexes was made by *Carrington, Schonland* and *Symons (71)*. The theoretical interpretation had to be modified when it was realized that the sub-shell energy difference Δ is, indeed, negative *(72, 73)*. We are not discussing here the d^1 and d^2 systems which still present intricate problems *(5, 6, 33)* but the tetroxo complexes of d^0 central atoms show a regular behaviour (the wave-number of the first strong electron transfer band is given in kK):

$$\text{Mn(VII)} < \text{Ru(VIII)} \sim \text{Cr(VI)} < \text{Os(VIII)} \sim \text{Tc(VII)} <$$
$$ \quad 18.6 \qquad 26.0 \qquad 26.8 \qquad 33.8 \qquad 34.6$$

$$\text{V(V)} < \text{Mo(VI)} \sim \text{Re(VII)} < \text{W(VI)}$$
$$36.9 \qquad 43.2 \qquad 43.7 \qquad 50.3 \tag{6}$$

rather in analogy to the variation of electron transfer spectra of hexahalides as a function of the central atom. *If* $x_{opt} = 3.0$ can be adopted for oxide Eq. (6) indicates a variation of x_{opt} from 2.4 for Mn(VII) to 1.3 for W(VI), more than half a unit below the values obtained for the hexahalides. *If* the oxo ligands have a *constant* x_{opt}, the optical electronegativity increases 0.3 unit for each unit of increasing oxidation number in an isoelectronic series, and it decreases 0.55 unit from the $3d$ to the $4d$ group and 0.3 unit from $4d$ to $5d$, under equal circumstances. Actually, Eq. (6) is very close to the difference between 99 kK and 8 kK times a quantity consisting of the sum of the oxidation number and 3 (in the $3d$ group), 1 (in the $4d$ group) or zero (in the $5d$ group). This agrees well with the chemically oxidizing character of the central atom, and once more, it is realized that oxides of a central atom having a given oxidation number are less oxidizing than hexahalide complexes.

If the situation was as simple as in the hexahalides *(6)* the tetrathio complexes should have their first strong electron transfer band at a constant difference below the wave-numbers in Eq. (6). Actually, the difference between VO_4^{-3} and VS_4^{-3} is 18.2 kK but between WO_4^{-2} and WS_4^{-2} 24.8 kK, and this difference is to a good approximation *(74, 75)* a *linear function* of Eq. (6). This is even true for the difference between tetrathio and tetraseleno complexes being 3.0 kK for vanadium, 3.4 kK for molybdenum and 3.9 kK for tungsten. It must be realized that some other observed energy differences are also, in part, linear functions of Eq. (6). For instance, the difference (in the following called Φ) of excitation energy between $(\pi)t_1$ and $(\pi)t_2$ is 13.6 kK for MnO_4^-, 12.4 kK for CrO_4^{-2}, 6 kK for RuO_4 and TcO_4^- and slightly below 5 kK for MoO_4^{-2} and ReO_4^- and it might be argued, like in the case of the second strong electron transfer band due to $(\pi)t_{2u}$ in hexahalides (6), that a more consistent set of optical electronegativities might be obtained from $(\pi)t_2$ in tetrahedral complexes.

It is not perfectly clear whether Φ is larger or smaller in the tetrathio and tetraseleno complexes than in the corresponding tetroxo complexes. Φ seems to be either 6.8 or 9.9 kK in VS_4^{-3} and possibly 5.9 kK in VSe_4^{-3} rather suggesting a decrease. As discussed below, several alternatives exist in the $4d$ and $5d$ groups, such as Φ in kK being:

	A	B	C
MoS_4^{-2}	small (2)	10.1	19.9
$MoSe_4^{-2}$	small (2.0)	9.8	19.2
WS_4^{-2}	small	10.6	20.7
WSe_4^{-2}	small(2.7)	10.0	19.7
ReS_4^-	small	12.2	24.2

$$(7)$$

The values in parentheses correspond to a weak shoulder *before* the first maximum. Though its distance is too large to be due to first-order spin-orbit coupling in the selenium atom, it may be the spin-forbidden transition to $^3T_1$ or a symmetry-forbidden transition (*i.e.* the "precursor band") to $^1T_1$ intensified by intermediate coupling and vibronic effects.

However, before we treat this problem, it may be worthwhile to mention the existence of tetrahedral complexes containing two or three different chalcogens as ligating atoms.

It is not only possible to prepare pure solids and solutions containing tetrathio and tetraseleno complexes, but also to exploit the highly different reaction rates (76) say, going from MO_4 over MO_3S, MO_2S_2, MOS_3 to MS_4 chromophores. In this way, stoichiometric mixed complexes can be isolated as crystalline salts. Further on, the formation of disproportionated mixed complexes takes from minutes to days in slightly alkaline solution, and it is possible to study the absorption spectra without too severe difficulties. The literature references to the individual complexes are: VS_4^{-3} ($77, 78$), VSe_4^{-3} ($75, 77$), $VO_2S_2^{-3}$ (79), VOS_3^{-3} (80), $VO_2Se_2^{-3}$ and $VOSe_3^{-3}$ (81), MoS_4^{-2} (82), $MoSe_4^{-2}$ ($83, 84$), MoO_3S^{-2} and $MoO_2S_2^{-2}$ (85), $MoOS_3^{-2}$ ($80, 86$), MoO_3Se^{-2} (87), $MoO_2Se_2^{-2}$ (84), $MoOSe_3^{-2}$ (88), $MoOS_2Se^{-2}$ (89), $MoOSSe_2^{-2}$ (90), MoS_3Se^{-2} and $MoSSe_3^{-2}$ ($91, 92$), WS_4^{-2} (82), WSe_4^{-2} ($83, 84$), WO_3S^{-2} and $WO_2S_2^{-2}$ (85), WOS_3^{-2} ($80, 86$), WO_3Se^{-2} (87), $WO_2Se_2^{-2}$ (84), $WOSe_3^{-2}$ (93), WOS_2Se^{-2} (89), $WOSSe_2^{-2}$ (90), ReS_4^- ($94, 95$), ReO_3S^- ($78, 79$) and $ReOS_3^-$ (79). The three ligating atoms O, S and Se allow the formation of twelve mixed tetrahedral complexes with a given central atom. Among the 48 conceivable mixed complexes of V, Mo, W and Re, half have been clearly characterized. For a short review, see (96) and for an extension of the spectra up to wave-numbers as high as 54 kK and a comparison with species such as CrO_4^{-2}, RuO_4 and OsO_4, see (97). Other tetroxo complexes recently discussed include VO_4^{-3} (77) and MnO_4^-, TcO_4^- and ReO_4^- (36).

Table 1 gives band positions (in kK) and molar extinction coefficients ε for d^0 complexes of four identical chalkogen atoms and the symmetry types of some of the excited states in the appropriate point-group T_d in the form of one-electron M.O. excitations among the orbitals of Eq. (4). Table 2 gives the same information about chromophores MX_3Y where the nuclei represent the point-group C_{3v} whereas Table 3 gives band positions and some ε values for MX_2Y_2 (point-group C_{2v}) and MX_2YZ (C_s only possessing a plane of symmetry). Certain unpublished results for free acids H_2WX_4 showing band splittings relative to WX_4^{-2} are included in Table 3.

The attempts of identifications of excited states fall in two categories: *induction* from an extensive comparison of chemically and structurally related complexes, as has, in general, been quite fruitful in the d-groups (7, 41) and *deduction* from approximative M.O. calculations. Unfortunately, the calculations are not always fully convincing for as many electrons as contained in our systems, and though it is beyond dispute that semi-empirical calculations frequently succeed in reproducing the assignments obtained by induction, the correct choice of assumptions is not always evident in the doubtful cases. Nevertheless, it can be very instructive to compare with semi-empirical results at various levels of sophistication, and in our case, the modification suggested by *Yeranos* of the older model of *Wolfsberg* and *Helmholz* has been applied to the nine MX_4 obtained by combining M = V, Mo, W with X = O, S, Se (98) and to TcO_4^-, ReO_4^- and ReS_4^- (99). Good agreement was obtained with the first electron transfer band identified as $(\pi)t_1 \rightarrow (d)e$, but the conclusions about the subsequent excited states are only tentative though interesting.

A comparison of Tables 1 and 2 shows the striking regularity (92) that the first band of MOS_3 almost coincides with the first band of MS_4 whereas the second band of MOS_3 almost coincides with the first band of MO_3S. These regularities are definitely not due to chemical impurities but can be rationalized in M.O. theory as a consequence of the exclusive $X(\pi)$ character of the a_2 component constituting a-third of the $(\pi)t_1$ orbitals of MX_4 when modified to MX_3Y. Said in other words, the same combination of sulphur orbitals producing the a_2 orbital of MOS_3 contributes to the set of three degenerate t_1 orbitals of MS_4. It is remarked in Table 2 that ε of this a_2 component is rather low, exactly as known from CrO_3X^- and MnO_3X with $X = OH$, F, Cl and Br. What is perhaps more unexpected is that MOS_3 (and not MO_3S) imitates this behaviour, as if X had a higher effective electronegativity than oxygen.

Table 1. *Wave-numbers in kK of maxima (or baricenters of vibrational structures) with shoulders in parentheses and molar extinction coefficients of complexes representing the point-group T_d. Identifications of M.O. symmetry type one-electron excitations are given for the first transitions*

	$(\pi)t_1 \rightarrow (d)e$	$(\pi)t_2 \rightarrow (d)e$	$(\pi)t_1 \rightarrow (d)t_2$	Other transitions		
VO_4^{3-}	36.9(8000)	45.0(5700)	—	—		
VS_4^{3-}	18.6(5800)	—	—	25.4(3500)	28.5(4800)	37.5(11000)
VSe_4^{3-}	15.6	—	—	21.5		
CrO_4^{2-}	(24),26.8(4800)	36.6(3700)	39.2(3000)	48.8(4700)		
MnO_4^-	(16),18.6(2400)	32.3(1700)	28.5(1200)	44.0(1000)	52.9(20200)	
MoO_4^{2-}	43.2(2800)	48.0(8400)	—	—		
MoS_4^{2-}	(19),21.4(13000)	—	—	31.5(17000)	41.3(24000)	48.3(40000)
$MoSe_4^{2-}$	(16),18.0	—	—	27.8	37.2	
TcO_4^-	34.6(2100)	40.5(5700)	—	(53.2)(1500)		
RuO_4	26.0	32.3	—	40(weak)		
WO_4^{2-}	50.3(8300)	—	—	—		
WS_4^{2-}	25.5(18500)	—	—	36.1(28500)	46.2(30000)	
WSe_4^{2-}	(18.9),21.6(16000)	—	—	31.6(24000)	41.3(30000)	
ReO_4^-	43.7(3600)	48.3(6100)	—	nextabove54		
ReS_4^-	19.8(9600)	—	—	32.0(18000)	44.0(30000)	
OsO_4	33.8	40.5	—	48(weak)		

When X and Y in MX_3Y do not present highly different electro-negativities (*41, 49*) the two degenerate M.O. forming the e component consist of $^5/_8$ $X(\pi)$ character and $^3/_8$ $Y(\pi)$ character in the squared ampli-tudes, restoring the baricenter rule for $(\pi)t_1$. In this limit, the energy difference between the a_2 and e component is three-eighths times the difference between the diagonal elements of energy of $X(\pi)$ and $Y(\pi)$. However, the e component is not alone when the π-orbitals are consider-ed in the point-group C_{3v} formed as a sub-group of T_d; the three $(\pi)t_2$ orbitals have the symmetry types a_1 and e in C_{3v} and the two $(\pi)e$ orbitals are still called e. In the opposite extreme of a large energy differ-ence between $X(\pi)$ and $Y(\pi)$, the first e component observed at the lowest wave-number is concentrated on the atoms X or Y having the least negative diagonal element of energy. The e components in Table 2 re-present an intermediate case between these two extremes, the observed distance 3 to 4 kK in MOS_3 and $MOSe_3$ (except 5.7 kK between a_2 and e of $ReOS_3^-$ being slightly more than one-eighth (rather than $^3/_8$) the

Table 2. *Absorption bands of complexes representing the point-group* C_{3v}. *Notation as in Table 1*

	$(\pi)a_2 \rightarrow (d)e$	$(\pi)e \rightarrow (d)e$	Other transitions		
VOS_3^{-3}	(19.2)	21.8	30.8	33.9	
$VOSe_3^{-3}$	15.6	19.2	—		
MoO_3S^{-2}	—	25.4	34.7(strong)	44.7	52.0
$MoOS_3^{-2}$	21.5(2300)	25.5(8700)	32.0(6600)(38.5)	44.1(14000)	
MoO_3Se^{-2}	—	22.0	32.0		
$MoOSe_3^{-2}$	17.9	22.0	28.45	(35.5)	40
MoS_3Se^{-2}	(18)(weak) and 20.6		30.6	40.4	
$MoSSe_3^{-2}$	(16)(weak) and 18.55		28.4	38.0	
WO_3S^{-2}	—	30.6	41.0(strong)		
WOS_3^{-2}	(26.7)(3000)	29.9(11300)	37.0(7200)	41.1(9700)	
WO_3Se^{-2}	—	27.0	38.0		
$WOSe_3^{-2}$	22.1	26.0(strong)	34.1	38.2	
ReO_3S^-	—	28.6	33.6(strong)	46.5	
$ReOS_3^-$	19.8	25.5	32.3		

distance between the first electron transfer band of MO_4 and MS_4 (or MSe_4).

It is seen from Table 3 that the mixed chromophores MO_2S_2 and MO_2Se_2 conserve a band at almost the same wave-number as the e component of the corresponding monoxo and trioxo complexes (*92*). This can be qualitatively understood as a comparable mixture with a major contribution from the less electronegative chalkogen. The M.O. description involves more free parameters in the point-group C_{2v} (thus, all five d-like orbitals can have different energies). On the other hand, mixed complexes containing sulphur and selenium show much less splitting of the T_d absorption bands and follow closely the rule of average environment (MS_nSe_{4-n} interpolating their wave-numbers as a linear function of n between MS_4 and MSe_4) as one would expect because of the minor difference (0.1 unit) of x_{opt} between thio and seleno ligands. The spectra of MOS_2Se and $MOSSe_2$ are rather similar to an extrapolation of MOX_3 involving one-third and two-thirds of sulphur or selenium.

As seen from Table 1, the effective x_{opt} is 0.6 to 0.8 units higher for oxide than for sulphide. This difference is not much smaller than between fluoride and chloride (0.9) or between water and iodide (1.0). Hence, already the presence of one ligating sulphur or selenium atom suffices to produce many absorption bands at lower wave-numbers than of the tetroxo complex. The three alternatives of Eq. (7) correspond to different interpretations of the bands above $(\pi)t_1$ (or above the a_2 and e components of Table 2).

Table 3. *Absorption bands of complexes representing the point-groups C_{2v} (and C_s and the two last probably C_1). Notation as in Table 1*

$VO_2S_2^{-3}$	21.8	27.8	32.8		
$VO_2Se_2^{-3}$	19.2 ?				
$MoO_2S_2^{-2}$	25.4(3000)	31.4(6000)	34.7(3000)		
$MoO_2Se_2^{-2}$	22.0	28.5	32.0		
$MoOS_2Se^{-2}$	(20)	24.3	31	42.5	
$MoOSSe_2^{-2}$	18.5	23.15	30.7		
$WO_2S_2^{-2}$	30.6(4000)	36.6(6900)	41.0(3900)		
$WO_2Se_2^{-2}$	27.0	34.0	38.0		
WOS_2Se^{-2}	(25)	29.1	36	41	48.4
$WOSSe_2^{-2}$	23.2	26.7	33.5		
H_2WS_4	(23.2)	26.1	36.0	39.1	
H_2WSe_4	(20.4)	23.0	32.1	35.6	
$H_2WO_2S_2$	(19.4)	23.7	33.8(strong)	37	43.7
$H_2WO_2Se_2$	(16.4)	(20.8)	30.5	35.7	40.5
H_2WOS_3	21.7	27.4(strong)	32.1	38.8(strong)	
H_2WOSe_3	17.6	24.5(strong)	27.9	35.3(strong)	

A: The sub-shell energy difference Δ between $(d)e$ and $(d)t_2$ (which we give here as a negative value in order to obtain a better concordance with the description of octahedral complexes) would be the number given in column C of Eq. (7) with opposite sign. A plausible interpretation of the number in column B would be the difference of excitation energy of $(\pi)t_2$ and $(\sigma)t_2$ like the difference 12 to 15 kK observed (4, 6) between $(\pi + \sigma)t_{1u}$ and $(\sigma + \pi)t_{1u}$ in hexahalides.

B: The number in column C multiplied by (-1) is also identified with Δ, but the number in column B is identified with Φ.

C: Conversely, the number in column B with opposite sign is Δ, whereas the number in column C represents Φ, the difference of excitation energy of $(\pi)t_1$ and $(\pi)t_2$. This choice was made in Ref. (97).

Unfortunately, we do not have a very clear-cut explanation of the variation of intensities of electron transfer bands. For instance, it is known that the intensities decrease with increasing oxidation number in an isoelectronic series of tetroxo complexes such as VO_4^{-3} to MnO_4^- or MoO_4^{-2} to RuO_4 and further on that the intensities increase, under equal circumstances, from $3d$ over $4d$ to $5d$. As a mnemotechnic device, one may remember that the oscillator strength of a transition is proportional to the square of an electric dipole moment, and that the larger tetrahedral complexes indeed have the higher intensities. However, the variation is far more pronounced than one would expect from the weak variation of the internuclear distances, and one can only conclude that the observed oscillator strength is derived from several, partly cancelling, contributions. In particular, one might have expected the $(\pi)t_1 \rightarrow (d)e$

transitions to be stronger when the internuclear distances are shorter and the covalent bonding stronger, whereas the opposite trend is observed in tetroxo complexes. In this connection, it is worthwhile to note that the three strong absorption bands of tetrathio and tetraseleno complexes with ε above 10000 are accompanied by a weak background with ε between 1000 and 2000 apparently containing broad bands, for instance close to 26 kK of MoS_4^{-2} and WSe_4^{-2} (97).

It is established beyond discussion (32) that Δ of the $4d^2$ system RuO_4^{-2} is more negative than -21 kK. This argument speaks against alternative C showing Δ values (97) in the tetrahedral $4d^0$ and $5d^0$ complexes between -10 and -13 kK, comparable to chromate and permanganate. On the other hand, Δ is between -19 and -25 kK in alternative B and conceivably more negative in alternative A, if the transitions to the upper sub-shell $(d)t_2$ are not ignored by being broad and weak. It is surprising enough that the number in column C of Eq. (7) for a given complex is so closely twice the number in column B. This would be compatible with a spreading dictated by group-theoretical conditions (41) as known from the *Hückel* model of benzene, but it is not evident that such a connection should exist between the numerical values, say of Φ and Δ. Our complexes do not, in general, present very strong bands such as the 52.9 kK band of MnO_4^- (36) except the band of MoS_4^{-2} about 27 kK above the first band. The corresponding transition most probably is $(\pi)t_2 \rightarrow (d)t_2$ in alternative B and $(\sigma)t_2 \rightarrow (d)t_2$ in alternative C, but it cannot be excluded that it is a $3p \rightarrow 4s$ excitation of the sulphur atom.

The fundamental question is whether Φ of tetrathio and tetraseleno complexes is much smaller than of the corresponding tetroxo complexes (A), slightly larger (B) or about three times as large (C). It is not easy to decide this question *a priori*; counteracting effects going from a smaller to a larger ligating atom are the longer internuclear distances decreasing the ligand-ligand repulsion and the stronger covalent bonding to the central atom. It may be that the chalkogenide complexes are roughly scaled, the radial extension of the ligands (represented by the reciprocal Slater exponents) changing proportional to the internuclear distances. As a matter of fact, the difference between the excitation energy of $(\pi)t_{1g}$ and $(\pi)t_{2u}$ having an origin comparable to that of Φ (41) is 10 kK in hexafluoro and 7 kK in hexachloro complexes (6) whereas the distance of diagonal elements (of *Russell-Saunders* coupling) are roughly 6 kK in hexabromo (49) and hexa-iodo complexes (57). This argument would favour alternatives A and B. It might be argued that Φ cannot have decreased from 5 to 6 kK in $4d$ and $5d$ group tetroxo complexes to so small a value for tetrathio complexes that alternative A would be acceptable. However, it is possible that $(\pi)t_2 \rightarrow (d)e$ in tetrathio and tetraseleno complexes have a considerably higher intensity than $(\pi)\,t_1 \to (d)\,e$ in which case the latter transition is not readily detected if Φ is 2 or 3 kK. The *Wolfsberg-Helmholz-Yeranos* calculations (*99*) give $\Phi = 12.3$ kK in TcO_4^- and 12.1 kK in ReO_4^- (roughly twice the observed values) and the increased $\Phi = 16.5$ kK in ReS_4^- which was taken (*97*) as an argument for alternative C, though it seems perhaps rather supporting B.

Jørgensen has the hunch that alternative A is the correct one, but it is also true that the vector sum of the many various arguments at present seems to point at alternative B.

5 Chalkogen-Bridging Ligands

One of the writers (*19*) suggested that nickel(II) forms brown solutions with $OAsS_2^{-3}$ because rectangular complexes $OAsS_2NiS_2AsO^{-4}$ containing the chromophore $Ni(II)\,S_4$ are so stable that NiS is not precipitated. This has never been positively confirmed, and unfortunately, the spectroscopic evidence is ambiguous since small concentrations of oligomeric complexes comparable to the hexameric mercaptide $Ni_6(SC_2H_5)_{12}$ produce strong brown colours with uncharacteristic broad spectra.

However, the corresponding $S_2WS_2NiS_2WS_2^{-2}$ has been characterized (*100, 101*). The strong absorption band of uncoordinated WS_4^{-2} at 25.5 kK (*cf.* also $(HS)_2WS_2$ (*121, 122*) in Table 3) is split into two strong bands at 23.8 and 26.3 kK (which occur in the corresponding $Zn(WS_4)_2^{-2}$ at 21.7 and 25.6 kK) whereas the singlet $\to$ singlet transitions in the $3d^8$ system $Ni(II)S_4$ are detected as shoulders at 14.3 and 19.0 kK at positions comparable to other low-spin quadratic complexes (*20*). $Co(WS_4)_2^{-2}$ probably contains the tetrahedral chromophore $Co(II)S_4$ having the strongest $3d^7$ quartet $\to$ quartet transition at 12.4 kK (*101*). It is also possible (*102*) to isolate salts of $Ni(MoS_4)_2^{-2}$ and $Zn(WSe_4)_2^{-2}$.

There is, of course, no sharp limit between this kind of complexes and bidentate ligands having two sulphur atoms bound both to the central atom and to a carbon or phosphorus atom such as dithiophosphates $(RO)_2PS_2^-$, dithiocarbamates $R_2NCS_2^-$ and xanthates $ROCS_2^-$ (*17, 20*). Recently, d-group complexes of $(C_6H_5)_2PS_2^-$ (*103, 104*), $(C_6H_5)_2PSe_2^-$ (*105*), $(C_6H_5)_2AsS_2^-$ (*106*) and $(C_6H_5)_2PSSe^-$ binding with one sulphur and one selenium atom (*107*) have been studied. The pronounced nephelauxetic effect of d^3 and d^6 central atoms demonstrates strong covalent bonding, as has also been pointed out by *Schäffer, Galsbøl* and *Furlani* in related complexes (*20*).

The thallium(I) salts of VS_4^{-3}, VSe_4^{-3}, NbS_4^{-3}, $NbSe_4^{-3}$, $MoSe_4^{-2}$, TaS_4^{-3}, $TaSe_4^{-3}$ and WSe_4^{-2} have been investigated (*108*). There is more interaction between the thallium and chalkogen atoms than in corresponding potassium or caesium salts, and hence, the chalkogens are bridging ligands. It is only possible to study the reflection (and not solution) spectra of these, highly coloured, solids, but some evidence is available for additional electron transfer bands at low wave-numbers due to transfer of a 6s electron from thallium(I) to the empty d-shell of the transition group atom in analogy to the additional electron transfer bands of silver(I) and thallium(I) salts of 5d group hexahalides (109).

The black colours of most d-group sulphides may be ascribed to collective empty orbitals to which transitions are also known in the visible of $TiCl_3$ and $TiBr_3$ and in the ultra-violet of many other halides (110). Only green Mn(II) S_6 and pink Mn(II)S_4 shows identifiable d^5 transitions in the crystalline modifications of MnS. Generally, the monomeric complexes of bidentate sulphur- and selenium-containing ligands (20) have far better characterized spectra. In solids, there is a very strong tendency for sulphur and selenium to bridge two or three atoms (70) with concomitant broad absorption bands in the visible.

6. Photo-Electron Spectra

It is of great importance for the M.O. description, in particular of diatomic and triatomic molecules, that the ionization energy I of penultimate M.O. below 21 eV can be experimentally determined (111) by photo-electron spectrometry of gaseous samples irradiated with monochromatic 21.2 eV photons. Though no volatile tetroxide has been studied (to our knowledge) it is important for our purposes to note that the "vertical" I (conserving the groundstate internuclear distances according to *Franck* and *Condon's* principle) of the three loosest bound M.O. of H_2O, *viz.* 12.6, 14.8 and 18.6 eV, show a greater spreading and higher values than 10.5, 13.3 and 15.4 eV for H_2S. We use here the unit of energy 1 eV $= 8.07$ kK.

The I values of solids can be obtained by irradiation with soft X-rays, such as 1486.6 eV photons (112). Unfortunately, the resolution is not as good, about 1 eV, and the quasi-stationary positive charge obtained by non-metallic samples can increase the apparent I values by 1 to 4 eV in typical cases. The valence region with I below 50 eV usually give rather weak signals, except when partly filled or completed d and f shells have low average radii. Nevertheless, *Prins* and *Novakov* (113) measured Li_2SO_4 and $LiClO_4$ (where the lithium 1s signal · occurs for I about 60 eV) and found I values in agreement with the order of filled orbitals of Eq. (4). Such studies have been continued (114) on related salts such as Li_2CO_3 and Li_3PO_4 and compared with M.O. calculations. The calculated value of Φ is 2.5 eV for SO_4^{-2} and 1.8 eV for ClO_4^- and the total spreading of the five highest sets of filled M.O. is calculated to be 10.8 and 13.5 eV for these two anions. Though the zero-point of energy is somewhat uncertain because of the charging effects, two weak signals of Li_2SO_4 at 8 and 10 eV can be ascribed to $(\pi)t_1$ and $(\pi)t_2$ whereas $(\pi)e$ is not clearly resolved. Two strong signals at $I = 15$ and 18 eV can be ascribed to $(\sigma)t_2$ and $(\sigma)a_1$, respectively. The corresponding I values in $LiClO_4$ seem to be 2 eV higher. As pointed out by *Manne et al.* (115), X-ray emission spectra can supplement valuable information to the photo-electron results, and in particular, the separation between $(\pi)t_2$ and $(\sigma)t_2$ in K_2SO_4 has been directly measured to be 3.5 eV. It is probable that these tetroxo complexes with unusually short internuclear distance have a wider spreading of M.O. energies than tetrathio complexes and even d^0 tetroxo complexes. *Jørgensen* and *Berthou* (116) discuss the influence of the cations on the valence region of sulphates. It must finally be noted that photo-electron spectra indicate "vertical" ionization energies and not, as electron transfer spectra (6) excitation energies influenced by charge separation effects.

The *inner shells* of atoms in compounds show chemical effects on I which are nearly the same for all the inner shells. This is what one would expect if the chemical shift indicates changes of the *Hartree* (including the *Madelung*) potential (8). Elements able to change their oxidation state by eight units, such as nitrogen, sulphur, and chlorine, are known to show chemical shifts amounting to 10 to 15 eV. However, even in the same oxidation state, *all* elements exhibit chemical shifts in intervals between 2 and 8 eV (117—119) though considerations of charging effects on isolating samples tend to diminuish these intervals by about a-third.

We have measured photo-electron spectra of many of the compounds discussed in this review (120) and find, in general, the lowest I values of a given central atom in the thio- and seleno complexes, as it is also true for other complexes (112, 119) of sulphur-containing ligands. Thus, the inner shells of rhenium in ReS_4^- have lower I than of various Re(IV) hexahalides, though it must be realized that also the considerably higher I values of ReO_4^- depend on the cation co-existing with the perrhenate. In the transition groups, the most important variable determining the chemical shifts seems to be the fractional charge being the highest in the fluoride and the least positive in the sulphide of a given central atom. However, measurements on potassium (I) show unexpected additional effects of instantaneous polarizability (117, 119).

A. M. and E. D. gratefully acknowledge the financial support of the *Deutsche Forschungsgemeinschaft.*

References

1. *Rabinowitch, E.:* Rev. Mod. Phys. *14,* 112 (1942).
2. *Linhard, M., Weigel, M.:* Z. Anorg. Allgem. Chem. *266,* 49 (1951).
3. *Fromherz, H.:* Z. Elektrochem. *37,* 553 (1931).
4. *Jørgensen, C. K.:* Mol. Phys. *2,* 309 (1959).
5. — Halogen Chemistry *1,* 265. London: Academic Press 1967.
6. — Progr. Inorg. Chem. *12,* 101 (1970).
7. — Oxidation Numbers and Oxidation States. Berlin–Heidelberg–New York: Springer 1969.
8. — Orbitals in Atoms and Molecules. London: Academic Press 1962.
9. *Barnes, J. C., Day, P.:* J. Chem. Soc. 3886 (1964).
10. *Schmidtke, H. H., Garthoff, D.:* J. Am. Chem. Soc. *89,* 1317 (1967).
11. *Jørgensen, C. K.:* Acta Chem. Scand. *16,* 2406 (1962).
12. *Abrahams, S. C., Ginsberg, A. P., Knox, K.:* Inorg. Chem. *3,* 558 (1964).
13. *Ryan, J. L., Jørgensen, C. K.:* Mol. Phys. *7,* 17 (1963).
14. *Jørgensen, C. K., Rittershaus, E.:* Mat. Fys. Medd. Danske Vidensk. Selskab *35,* no. 15 (1967).
15. *Bergerhoff, G.* and *Paeslack, J.:* Z. Krist. *126,* 112 (1968).
16. *Caro, P.:* J. Less-Common Metals *16,* 367 (1968).
17. *Zollweg, R. J.:* Phys. Rev. *111,* 113 (1958).
18. *Moreau-Colin, M. L.:* J. Chim. Phys. *67,* 498 (1970); Struct. Bonding *10,* 167 (1972).
19. *Jørgensen, C. K.:* J. Inorg. Nucl. Chem. *24,* 1571 (1962).
20. — Inorg. Chim. Acta Rev. (Padova) *2,* 65 (1968).
21. — *Pappalardo, R., Flahaut, J.:* J. Chim. Phys. *62,* 444 (1965).
22. *Schmidtke, H. H.:* Ber. Bunsenges. Physik. Chem. *71,* 1138 (1967).
23. *Jensen, K. A., Krishnan, V., Jørgensen, C. K.:* Acta Chem. Scand. *24,* 743 (1970).
24. *Orgel, L. E.:* Mol. Phys. *7,* 397 (1964).
25. *Jørgensen, C. K.:* Advan. Chem. Phys. *5,* 33 (1963).
26. *Wolfsberg, M., Helmholz, L.:* J. Chem. Phys. *20,* 837 (1952).
27. *Jørgensen, C. K., Horner, S. M., Hatfield, W. E., Tyree, S. Y.:* Int. J. Quantum Chem. *1,* 191 (1967).
28. *Carrington, A., Symons, M. C. R.:* Chem. Rev. *63,* 443 (1963).
29. *Brendel, C., Klemm, W.:* Z. Anorg. Allgem. Chem. *320,* 59 (1963).
30. *Gray, H. B.:* Coordin. Chem. Rev. *1,* 2 (1966).
31. *De Michelis, G., Oleari, L., Di Sipio, L., Tondello, E.:* Coordin. Chem. Rev. *2,* 53 (1967).
32. *Carrington, A., Jørgensen, C. K.:* Mol. Phys. *4,* 395 (1961).
33. *Day, P., Di Sipio, L., Oleari, L.:* Chem. Phys. Letters *5,* 533 (1970).
34. *Wells, E. J., Jordan, A. D., Alderdice, D. S., Ross, I. G.:* Australian J. Chem. *20,* 2315 (1967).
35. *Langseth, A., Qviller, B.:* Z. Physik. Chem. B *27,* 79 (1934).
36. *Mullen, P., Schwochau, K., Jørgensen, C. K.:* Chem. Phys. Letters *3,* 49 (1969).
37. *Jørgensen, C. K.:* Acta Chem. Scand. *17,* 1043 (1963).
38. *Holt, S. L., Ballhausen, C. J.:* Theoret. Chim. Acta *7,* 313 (1967).
39. *Schatz, P. N., McCaffery, A. J., Suëtaka, W., Henning, G. N., Ritchie, A. B., Stephens, P. J.:* J. Chem. Phys. *45,* 722 (1966).
40. — *McCaffery, A. J.:* Quart. Rev. (London) *23,* 552 (1969).
41. *Jørgensen, C. K.:* Modern Aspects of Ligand Field Theory. Amsterdam: North-Holland Publishing Co. 1971.

42. *Henning, G. N., McCaffery, A. J., Schatz, P. N., Stephens, P. J.:* J. Chem. Phys. *48*, 5656 (1968).
43. *McCaffery, A. J., Schatz, P. N., Lester, T. E.:* J. Chem. Phys. *50*, 379 (1969).
44. *Bird, B. D., Day, P., Grant, E. A.:* J. Chem. Soc. (A) 100 (1970).
45. *Briat, B., Rivoal, J. C., Petit, R. H.:* J. Chim. Phys. *67*, 463 (1970).
46. *DiSipio, L., De Michelis, G., Tondello, E., Oleari, L.:* Gazz. Chim. Ital. *96*, 1785 (1966).
47. *Brisdon, B. J., Lester, T. E., Walton, R. A.:* Spectrochim. Acta *23 A*, 1969 (1967).
48. *Ryan, J. L.:* Inorg. Chem. *8*, 2058 (1969).
49. *Jørgensen, C. K., Preetz, W.:* Z. Naturforsch. *22 a*, 945 (1967).
50. — Mol. Phys. *5*, 271 (1962).
51. *Ryan, J. L., Jørgensen, C. K.:* J. Phys. Chem. *70*, 2845 (1966).
52. *Jørgensen, C. K.:* Mol. Phys. *6*, 43 (1963).
53. *Day, P., Jørgensen, C. K.:* J. Chem. Soc. 6226 (1964).
54. *Bird, B. D., Day, P.:* J. Chem. Phys. *49*, 392 (1968).
55. — — Chem. Comm. (London) 741 (1967).
56. *Preetz, W., Homborg, H.:* J. Inorg. Nucl. Chem. *32*, 1979 (1970).
57. *Jørgensen, C. K., Preetz, W., Homborg, H.:* Inorg. Chim. Acta (Padova) *5*, 223 (1971).
58. *Helmholz, L., Brennan, H., Wolfsberg, M.:* J. Chem. Phys. *23*, 853 (1955).
59. *Jørgensen, C. K.:* Acta Chem. Scand. *11*, 73 (1957).
60. *Murmann, R. K., Foerster, D. R.:* J. Phys. Chem. *67*, 1383 (1963).
61. *Glemser, O., Roesky, H., Hellberg, K. H.:* Angew. Chem. *75*, 346 (1963).
62. *Krauss, H. L., Stark, K.:* Z. Naturforsch. *17 b*, 1 (1962).
63. *Ahlborn, E., Diemann, E., Müller, A.:* Z. Naturforsch. *27 b*. 1108 (1972).
64. *Butowiez, B.:* Compt. Rend. (Paris) B *266*, 1083 (1968).
65. *Aymonino, P. J., Schulze, H., Müller, A.:* Z. Naturforsch. *24 b*, 1508 (1969).
66. *Briggs, T. S.:* J. Inorg. Nucl. Chem. *30*, 2866 (1968).
67. *Royer, D. J.:* J. Inorg. Nucl. Chem. *17*, 159 (1961).
68. *Kläning, U.:* Acta Chem. Scand. *11*, 1313 (1957).
69. *Müller, A., Baran, E. J., Bollmann, F., Aymonino, P. J.:* Z. Naturforsch. *24 b*, 960 (1969).
70. *Jørgensen, C. K.:* Inorganic Complexes. London: Academic Press, 1963.
71. *Carrington, A., Schonland, D., Symons, M. C. R.:* J. Chem. Soc. 4710 (1956) and 659 (1957).
72. *Lott, K. A. K., Symons, M. C. R.:* J. Chem. Soc. 829 (1959).
73. *Carrington, A., Symons, M. C. R.:* J. Chem. Soc. 889 (1960).
74. *Müller, A., Krebs, B., Glemser, O., Diemann, E.:* Z. Naturforsch. *22 b*, 1235 (1967).
75. — *Diemann, E.:* Z. Chem. *8*, 197 (1968).
76. *Aymonino, P. J., Ranade, A. C., Diemann, E., Müller, A.:* Z. Anorg. Allgem. Chem. *371*, 300 (1969).
77. *Müller, A., Diemann, E., Ranade, A. C.:* Chem. Phys. Letters *3*, 467 (1969).
78. — *Krebs, B., Rittner, W., Stockburger, M.:* Ber. Bunsenges. Physik. Chem. *71*, 182 (1967).
79. *Ranade, A. C., Müller, A., Diemann, E.:* Z. Anorg. Allgem. Chem. *373*, 258 (1970).
80. *Diemann, E., Müller, A.:* Spectrochim. Acta *26 A*, 215 (1970).
81. *Müller, A., Ranade, A. C., Rao, V. V. K.:* Spectrochim. Acta *27 A*, 1973 (1971).
82. — *Rittner, W., Nagarajan, G.:* Z. Physik. Chem. *54*, 229 (1967).

83. — *Krebs, B., Glemser, O., Diemann, E.*: Z. Naturforsch. *22 b*, 1235 (1967).
84. — *Diemann, E.*: Chem. Ber. *102*, 945 (1969).
85. — — *Ranade, A. C., Aymonino, P. J.*: Z. Naturforsch. *24 b*, 1247 (1969).
86. — — *Heidborn, U.*: Z. Anorg. Allgem. Chem. *371*, 136 (1969).
87. — *Ranade, A. C., Rittner, W.*: Z. Anorg. Allgem. Chem. *380*, 76 (1971).
88. — *Diemann, E.*: Chem. Ber. *102*, 3277 (1969).
89. — — Chem. Ber. *102*, 2603 (1969).
90. — — Z. Anorg. Allgem. Chem. *373*, 57 (1970).
91. — *Menge, R., Neumann, F.*: Z. Anorg. Allgem. Chem., submitted.
92. — *Diemann, E., Neumann, F., Menge, R.*: Chem. Phys. Letters *16*, 521 (1972).
93. — *Diemann, E.*: Chem. Ber. *102*, 2044 (1969).
94. — *Krebs, B., Diemann, E.*: Z. Anorg. Allgem. Chem. *353*, 259 (1967).
95. — *Diemann, E., Rao, V. V. K.*: Chem. Ber. *103*, 2961 (1970).
96. — Chimia (Aarau) *24*, 346 (1970).
97. — *Diemann, E.*: Chem. Phys. Letters *9*, 369 (1971).
98. *Kebabcioglu, R., Müller, A.*: Chem. Phys. Letters, *8* 59 (1971).
99. — — *Rittner, W.*: J. Mol. Struct. *9*, 207 (1971).
100. *Müller, A., Diemann, E.*: Chem. Comm. (London) 65 (1971).
101. — — *Heinsen, H. H.*: Chem. Ber. *104*, 975 (1971).
102. — *Ahlborn, E., Heinsen, H. H.*: Z. Anorg. Allgem. Chem. *386*, 102 (1971).
103. — *Rao, V. V. K., Diemann, E.*: Chem. Ber. *104*, 461 (1971).
104. — — *Klinksiek, G.*: Chem. Ber. *104*, 1892 (1971).
105. — *Christophliemk, P., Rao, V. V. K.*: Chem. Ber. *104*, 1905, (1971).
106. — *Werle, P.*: Chem. Ber. *104*, 3782 (1971).
107. — *Rao, V. V. K., Christophliemk, P.*: J. Inorg. Nucl. Chem. *34*, 345 (1972).
108. — *Schmidt, K. H., Tytko, K. H., Bouwma, J., Jellinek, F.*: Spectrochim. Acta *28 A*, 381 (1972).
   Omloo, W. P. F. A. M., Jellinek, F., Müller, A., Diemann, E.: Z. Naturforsch. *25b*, 1302 (1970).
109. *Jørgensen, C. K.*: Acta Chem. Scand. *17*, 1034 (1963).
110. *Clark, R. J. H.*: J. Chem. Soc. 417 (1964).
111. *Turner, D. W., Baker, C., Baker, A. D., Brundle, C. R.*: Molecular Photoelectron Spectroscopy. London: Wiley-Interscience 1970.
112. *Jørgensen, C. K.*: Chimia (Aarau) *25*, 213 (1971); *26*, 252 (1972).
113. *Prins, R., Novakov, T.*: Chem. Phys. Letters *9*, 593 (1971).
114. *Connor, J. A., Hillier, I. H., Saunders, V. R., Barber, M.*: Mol. Phys. *23*, 81 (1972).
115. *Manne, R., Karras, M., Suoninen, E.*: Chem. Phys. Letters *15*, 34 (1972).
116. *Jørgensen, C. K., Berthou, H.*: Helv. Chim. Acta, under preparation.
117. — — *Balsenc, L.*: J. Fluorine Chem. *1*, 327 (1972).
118. — Theoret. Chim. Acta *24*, 241 (1972).
119. — *Berthou, H.*: Mat. Fys. Medd. Danske Vidansk. Selskab *38*, no. 15 (1972).
120. *Müller, A., Jørgensen, C. K., Diemann, E.*: Z. Anorg. Allgem. Chem. *391*, 38 (1972).
121. *Müller, A., Schulze, H.*, unpublished results, cf. *Diemann, E., Müller. A.*, Coord. Chem. Rev., *10*, 79 (1973).
122. *Gattow, G., Franke, A.*: Z. Anorg. Allgem. Chem. *352*, 11, 246 (1967).

Received August 22, 1972

The Evidence for "Out-of-the-Plane" Bonding in Axial Complexes of the Copper(II) Ion

B. J. Hathaway

The Chemistry Department, University College, Cork, Ireland

Table of Contents

As the number of axial copper(II) complexes of known crystal structure continues to increase (1, 2), the "effective" coordination number of the local copper(II) ion environment is becoming less certain. In many recent textbooks (3, 4) the stereochemistry of the copper(II) ion is described as being dominated by the four coordinate square-coplanar stereochemistry, involving four relatively short in-plane bonds (R_s) of ca. 2.0 Å (Fig. 1). The presence of further ligands along the axial directions, at appreciably longer bond-lengths, R_L (where $R_L - R_s \approx$ 0.6 Å) is recognised (5) and described by such expressions as $4 + 2$ coordination (elongated tetragonal octahedral) or $4 + 1$ coordination (square pyramidal). However, it is never made clear whether or not these fifth and sixth axial ligands are actually involved in coordination along these axial directions, or are simply present due to the packing of the complex

49

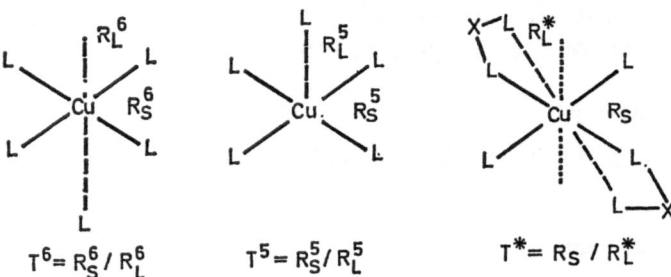

Fig. 1. The definition of tetragonality (T) for axial copper(II) complexes. The super script-number refers to the effective coordination number of the copper(II) ion and will be omitted if the coordination number is unimportant or not obvious

* refers to the ligand off-the-z-axis

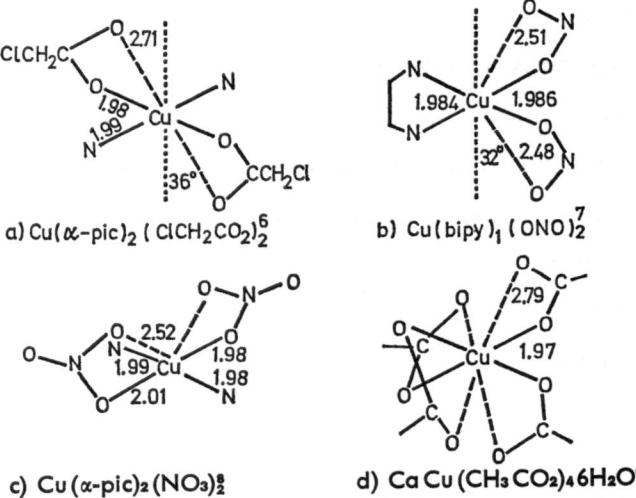

a) $Cu(\alpha\text{-pic})_2 (ClCH_2CO_2)_2^6$

b) $Cu(bipy)_1 (ONO)_2^7$

c) $Cu(\alpha\text{-pic})_2 (NO_3)_2^8$

d) $CaCu(CH_3CO_2)_4 6H_2O^9$

Fig. 2. Examples of copper(II) stereochemistries involving out-of-the-plane ligands positioned off-the-z-axis

in the crystal lattice. More recently, structures have been determined involving potential ligand atoms positioned both out of the xy plane and off-the-z-axis (6—9) (Fig. 2) or in some complexes the structures involve a mixture of off- and along-the-z-axis ligands (10, 11) Fig. 3). The re-cognition of such structures, further extends the question of whether or not such ligands are involved in coordination to the copper(II) ion and, further, whether a strictly square-coplanar CuL_4 chromophore involving four sigma-bonding ligands can have a separate existence.

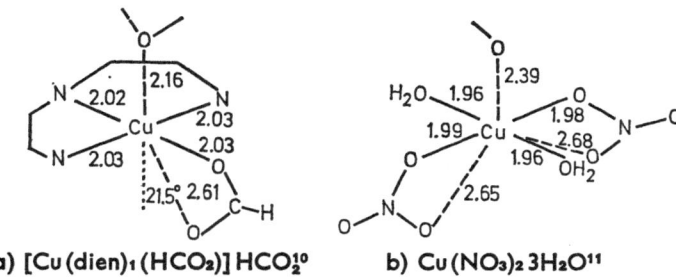

a) $[Cu(dien)_1(HCO_2)]HCO_2^{10}$ b) $Cu(NO_3)_2 3H_2O^{11}$

Fig. 3. Examples of copper(II) stereochemistries involving out-of-the-plane ligands positioned both along, and off-the-z-axis

It therefore seems appropriate to collect together the different types of evidence for out-of-the-plane bonding of ligands in axial copper-(II) complexes and to assess the different methods available for satisfying the out-of-the-plane bonding potential of a strictly square coplanar CuL_4 chromophore.

A. Crystallographic Data

Table 1 summarises (12) the range of observed copper-ligand bond-lengths for a number of different donor ligand atoms in axial stereochem-istries. The average values of R_L clearly parallel the average values of R_S, both increasing with the size of the covalent radii of the ligand donor

Table 1. *Copper-ligand bond-lengths, R_S, R_L, R_L-R_S, R_S/R_L and covalent radii of ligand atoms (Å)*

Ligand atom	R_S Range	Mean	R_L Range	Mean	R_L-R_S	R_S/R_L	Ligand atom covalent radii
F	1.89—1.93	1.91(10)	2,21—2.86	2.36(12)	0.45	0.809	0.64
O[a]	1.92—2.16	1.99(42)	2.22—2.89	2.50(16)	0.51	0.796	0.66
N[b]	1.99—2.14	2.03(40)	—	—	—	—	0,70
Cl	2.25—2.34	2.31(48)	2.73—3.19	2.93(32)	0.62	0,792	0.99
Br	2.40—2.56	2.45(12)	3.08—3.19	3.15(6)	0.70	0.777	1.14

[a]) Water   [b]) Ammonia

atom (3). The values of $R_L - R_S$ also increase with increasing size of the donor atom, while the ratio of R_S/R_L (referred to as the tetragonality T (12), Table 1) is essentially constant at 0.81—0.78 and suggests that these axial ligands are present at a definite out-of-the-plane distance, a structural situation which has been described (2, 12) as "semi-coordination". Comparable out-of-the plane distances (6—11) also occur in complexes involving ligands positioned off-the-z-axis, (Fig. 2 and 3, and Table 2). When off-the-z-axis ligands occur there is a comparability not

Table 2. *Copper-oxygen bond-lengths R_S and R_L^* involving off-z-axis ligands, $R_L^*-R_S$ and $T^* = R_S/R_L^*$*

Coordination	Complex	R_S	R_L^*	$R_L^*-R_S$	T^*
6	$Cu(NH_3)_2(CH_3CO_2)_2$	2.07	2.77	0.70	0.749
	$Cu(pyrazine)(NO_3)_2$	2.01	2.49	0.48	0.780
	$Cu(\alpha\text{-pic})_2(NO_3)_2$ I	1.99[a]	2.40[a]	0.41	0.829
	$Cu(\alpha\text{-pic})_2(NO_3)_2$ II	1.99	2.534	0.54	0.787
	$Cu(\alpha\text{-pic})_2(ClCH_2CO_2)_2$	1.98	2.71	0.73	0.73
	$Cu(bipy)_1(ONO)_2$	1.986[a]	2.49[a]	0.50	0.797
8	$CaCu(CH_3CO_2)_4\ 6\ H_2O$	1.97	2.79	0.82	0.706
	$Cu(H_4A_4)(ClO_4)_2$	1.93	2.88	0.95	0.67

[a]) mean values of two.

only in the out-of-the-plane bond-length $^*R_L$ with that involved in along the z-axis coordination, such that $^*R_L \approx R_L$ and $^*R_S \approx R_S$, but there is also a corresponding parallel with change of coordination number. Thus, while the out-of-plane bond-lengths involved in a six coordinate complex (R_L^6) are longer than those for a five coordinate complex (R_L^5) (Table 3), the corresponding bond-lengths in off-the-z-axis coordination ($^*R_L^6$ and $^*R_L^5$) are not only comparable, but $^*R_L^6 > ^*R_L^5$ (see Table 2).

Table 3. *Square pyramidal copper(II) complexes involving oxygen or nitrogen ligands, (mean of 14 complexes)*

Mean R_S^5	Mean R_L^5	$R_L^5-R_S^5$	$R_S^5/R_L^5 = (T^5)$
1.98	2.38	0.40	0.833

With the presence of off-the-z-axis coordination restricted to trigonal planar ligands such as nitrate and carboxylate ions it could be argued that the out-of-plane ligands are held close ot the copper(II) ion by the geometry of these polyanions. However, firstly there is no reason why the terminal oxygen ligands should not be positioned remote from the copper(II) ion (Fig. 4) and still retain the 120° Cu—O—N angle, and

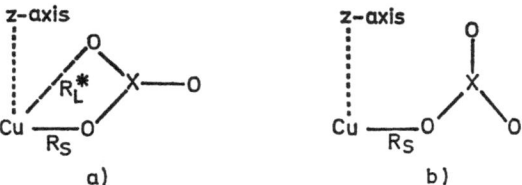

Fig. 4. The alternative conformation of the terminal oxygen atom for trigonal molecules (XO$_3$) coordinated to the copper(II) ion in the plane a) Out-of-the-plane bonding, b) No out-of-the-plane bonding

secondly the above bond-length correlation is difficult to understand unless these axial ligands occupy definite structural positions. Consequently the addition of the various possible axial and off-the-z-axis types of coordination to a basic square-coplanar CuL$_4$ chromophore can give rise to a range of stereochemistries involving coordination numbers from five to eight (Table 4).

Table 4. *Variable Coordination Number*

Coordination No.	Types of Bond lengths	Centre of Symmetry	Example
Five	4 + 1	No	[Cu(NH$_3$)$_4$H$_2$O]SO$_4$
Six	4 + 2	Yes	Cu(NH$_4$)$_3$(NO$_2$)$_2$
	4 + 1 + 1*)	No	[Cudien(HCO$_2$)]HCO$_2$
	4 + 0 + 2*)	Yes	Cu(α-pic)$_2$(ClCH$_2$CO$_2$)$_2$
	4 + 0 + 2*)	No	Cu(α-pic)$_2$(NO$_3$)$_2$
Seven	4 + 1 + 2*)	No	Cu(NO$_3$)$_2$2.5 H$_2$O
Eight	4 + 0 + 4*)	No	CaCu(CH$_3$CO$_2$)$_4$6 H$_2$O
		No	Cu(H$_4$A$_4$) (ClO$_4$)$_2$

*) refers to off-the-z-axis ligand atoms.

B. Infrared Spectra of Polyatomic Anions

Some evidence for weak coordination (semi-coordination) of axial ligands can be obtained (12) from the infrared spectra of polyatomic anions when these anions are involved in the axial positions of an elongated tetragonal octahedral stereochemistry. Fig. 5 shows two such examples for the perchlorate ion and Fig. 6 shows (14) the infrared spectrum of the $Cu(NH_3)_4(ClO_4)_2$ complex. The free perchlorate ion (15), as in $KClO_4$, gives rise to two infrared bands in this region; ν_1 the symmetric stretching mode at 920 cm^{-1}, which is strictly infrared forbidden and ν_3 the assymetric bending mode at ca. 1100 cm^{-1} which is infrared allowed. When the perchlorate ion is involved in semi-coordination (12) its sym-

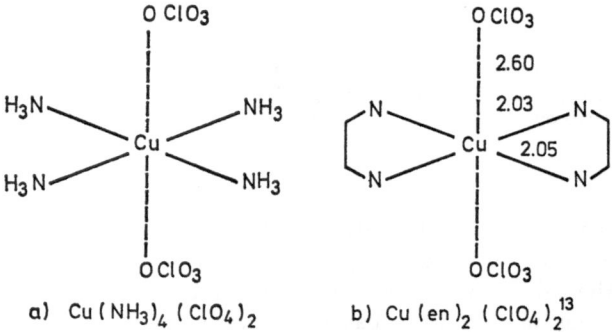

a) $Cu(NH_3)_4(ClO_4)_2$ b) $Cu(en)_2(ClO_4)_2$[13]

Fig. 5. The molecular structures of (a) $Cu(NH_3)_4(ClO_4)_2$ and (b) $Cuen_2(ClO_4)_2$

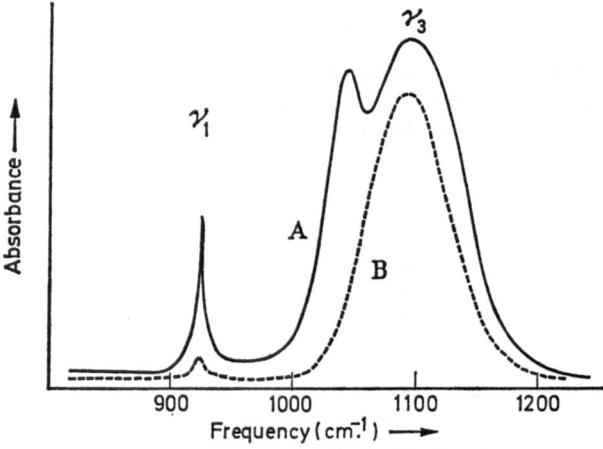

Fig. 6. The infrared spectra of (a) $Cu(NH_3)_4(ClO_4)_2$; and (b) $KClO_4$

metry is lowered from T_d to C_{3v} in which symmetry ν_1 becomes infrared allowed and the triple degeneracy of ν_3 is partly removed. Consequently in the infrared spectrum of $Cu(NH_3)_4(ClO_4)_2$ the band at 920 cm^{-1} becomes more intense (30—40%) and the band at 1100 cm^{-1} is resolved into two bands (Fig. 6) and clearly suggests some small but positiove interaction between these axial perchlorate groups and the copper(II) ions present. Comparable infrared evidence for semi-coordination has been obtained for the sulphate (16) and tetrafluoroborate (17) anions in elongated tetragonal octahedral copper(II) complexes. Fig. 7 shows

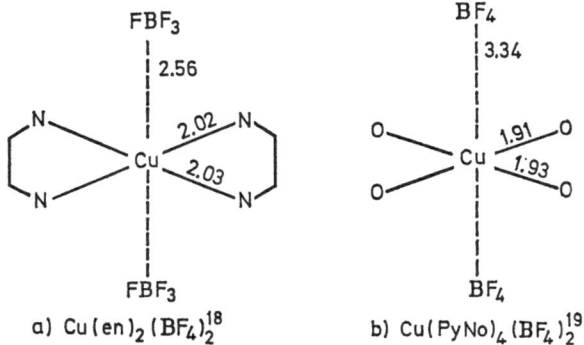

Fig. 7. The molecular structures of (a) Cuen$_2$ (BF$_4$)$_2$ and (b) Cu(PyNO)$_4$(BF$_4$)$_2$

the structure of two axial tetrafluoroborate complexes with significantly different values of R_L. In Cuen$_2$(BF$_4$)$_2$ the infrared spectrum gives clear evidence of semi-coordinated tetrafluoroborate ions, while in Cu(Py-NO)$_4$(BF$_4$)$_2$ the infrared spectrum shows the presence of ionic tetrafluoroborate ions. The reason for this difference lies in the nature of the four in-plane ligands; the nitrogens of ethylenediamine can only act as a σ-bonding ligand while the oxygen of pyridine N-oxide can act as a σ-and π-bonding ligand, the latter function of which can satisfy the additional out-of-the-plane bonding potential of the square coplanar CuO_4 chromophore by π-bonding (see later).

C. Electronic Energy Levels

Further evidence for the interaction of these axial ligands may be obtained from the way in which the electronic energy levels (2) of the central copper(II) ion varies with the tetragonality present (R_S/R_L), Fig. 8.

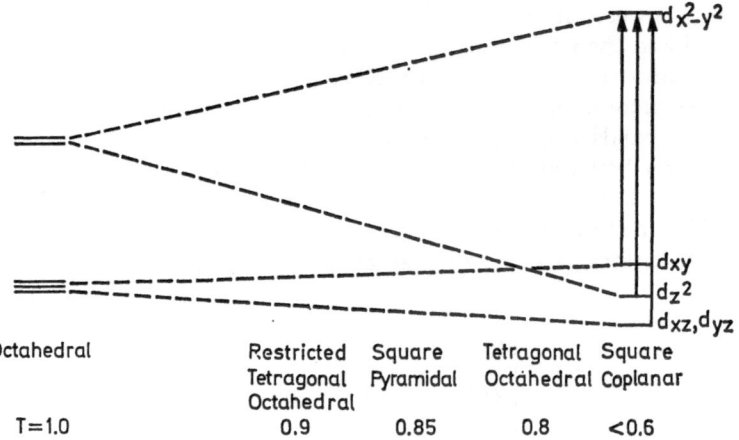

Fig. 8. The electronic energy levels of elongated axial copper(II) complexes

The energy of the $d_{z^2} \rightarrow d_{x^2-y^2}$ transition is most sensitive to the presence of axial ligands which interact directly with the d_{z^2} orbital on the copper(II) ion. Such a correlation diagram has been established using the technique of polarised single-crystal electronic spectroscopy (20) which not only resolves the component bands of the three possible transitions of Fig. 8 as in Fig. 9, but under favourable circumstances may allow the

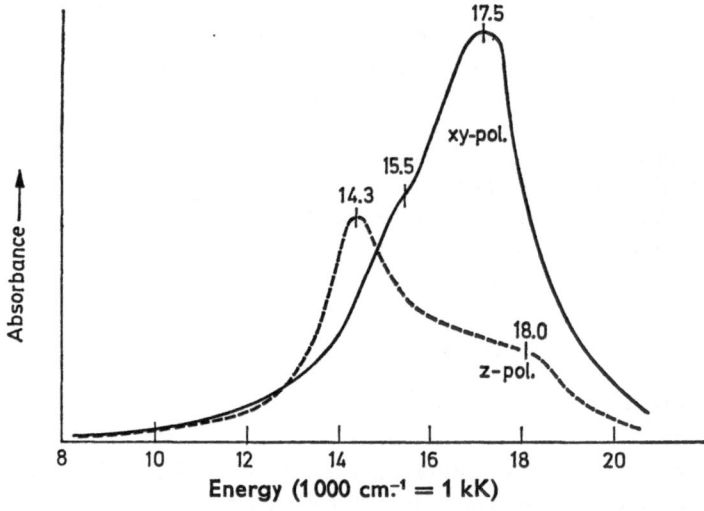

Fig. 9. The polarised single-crystal electronic spectra of $Cu(NH_3)_4(NO_2)_2$

absolute assignment of these transitions. The sensitivity of the energy of the $d_{z^2} \rightarrow d_{x^2-y^2}$ transition to the stereochemistry present is illustrated (24—26) for the three complexes of Fig. 10, in Fig. 11, in which the energies of the $d_{z^2} \rightarrow d_{x^2-y^2}$ transition in each complex are indicated by an

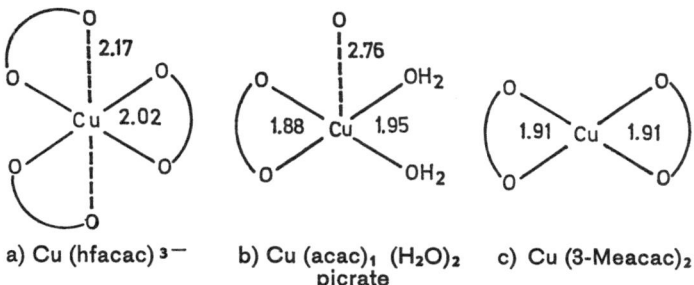

a) Cu (hfacac) $^{3-}$ b) Cu (acac)$_1$ (H$_2$O)$_2$ c) Cu (3-Meacac)$_2$
                                    picrate

Fig. 10. The molecular structures of (a) Cu(hfacac)$^{3-}$; (b) Cu(acac)$_1$(H$_2$O)$_2$ picrate; (c) Cu(3-Meacac)$_2$

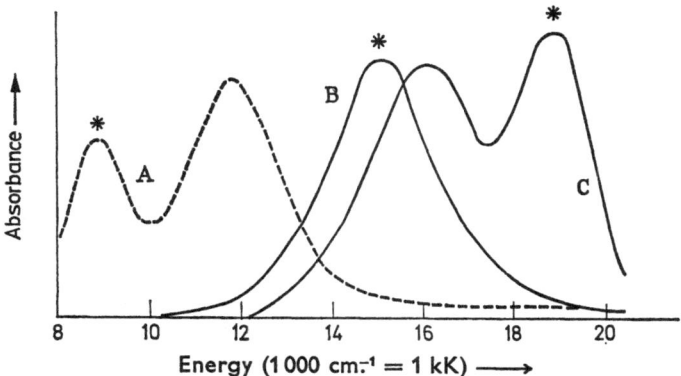

Fig. 11. The electronic reflectance spectra of (a) Cu(hfacac)$^{3-}$; (b) Cu(acac)$_1$(H$_2$O)$_2$ picrate; (c) Cu(3-Meacac)$_2$

asterisk and show a total range of ca. 12.0 kK between the different stereochemistries involved. Unfortunately the energy of the $d_{z^2} \rightarrow d_{x^2-y^2}$ transition is not just a function of the out-of-plane bond-length it also varies with the value of R_S; the shorter R_S the higher the energy. This is shown in Fig. 12, where the more rapid variation of R_L than that of R_S, with tetragonality does suggest that the former is a more sensitive criterion.

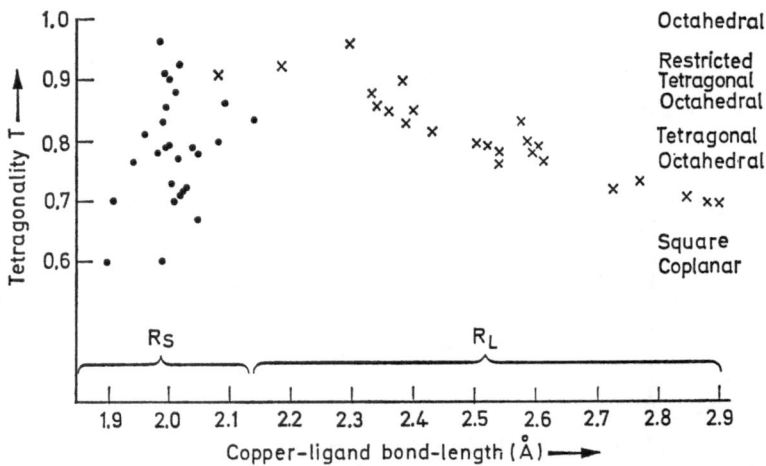

Fig. 12. The variation of the copper-ligand bond-lengths in-plane (R_S) and out-of-plane (R_L) for elongated octahedral copper(II) complexes (oxygen and nitrogen donor atoms only) vs. tetragonality (T)

D. Copper-Ligand Infrared Spectra

The differences in the values of R_S and R_L clearly parallel the tetragonality present (Fig. 12) and should be reflected in the frequency of the copper-ligand asymmetric stretching mode of vibration. Data for the in-plane copper-nitrogen modes has been obtained for a series of axial copper(II) ammines of varied stereochemistries (27). If the frequency of the Cu—N mode is plotted against the band maxima of the electronic spectra, the results of Fig. 13 are obtained. A significant correlation exists and it would be invaluable to obtain a corresponding comparison between the infrared frequency of the axial copper-ligand bonds and the maxima of the electronic spectra. Unfortunately the frequency of this bond, due to its greater length and hence presumably lower strength lies to significantly lower energy, probably <200 cm^{-1} for copper-nitrogen bonds. No clear assignment of the infrared spectra in this region has been reported. A detailed polarised infrared spectrum and Raman spectrum of the $CuO_2Cl_2Cl_2$ chromophore (28) in $CuCl_2 2 H_2O$ and $K_2Cu(H_2O)_2Cl_4$ have been reported and the frequency of the long copper-chlorine bond has been assigned. Unfortunately the infrared and Raman spectra do not agree and in any case whether or not the copper(II) environment is described as square-coplanar or elongated tetragonal octahedral is not determined by the experimental data and has to be assumed before the spectra can be assigned. Consequently these results do not, at present, give any information on the existence of an axial copper-ligand bond.

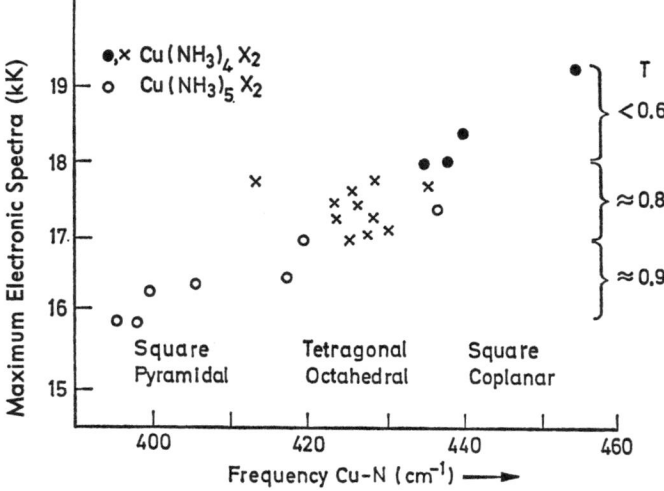

Fig. 13. The electronic specta (band maximum) vs. Cu—N frequency of copper(II) tetraammines and pentaammines

E. Square Pyramidal Stereochemistry

The effect of a single fifth ligand at a distance R_L^5 along the z-axis of a square-coplanar chromophore will clearly have less effect on the electronic energy levels, particularly the energy of the $d_{z^2} \rightarrow d_{x^2-y^2}$ transition, than the effect of two ligands at comparable distances, one above and one below the plane of the CuL_4 chromophore. Nevertheless, as the out-of-plane bond-lengths observed in square pyramidal coppe1(II) complexes (1, 2) are usually significantly shorter than those involved in an elongated tetragonal octahedral stereochemistry (Tables 1 and 3) the apparent tetragonalities T^5 are higher (Table 3) but they do not necessarily produce lower energies for the $d_{z^2} \rightarrow d_{x^2-y^2}$ transition (Table 6) than those observed for the elongated tetragonal octahedral stereochemistry. At R_L^5 bond-lengths of 2.1—2.3 Å there can be no reasonable doubt that appreciable bonding must occur. Consequently, as two semi-coordinate ligands at 2.6 Å produce comparable values of the energy of the $d_{z^2} \rightarrow$

$d_{x^2-y^2}$ transition, this suggests a comparable amount of interaction between these two ligands and the d_{z^2} orbital, consistent with weak bonding.

Table 5. *The energy of the $d_{z^2} \to d_{x^2-y^2}$ transition in square pyramidal copper(II) complexes as a function of R_L^5 and T^5*

Complex	Energy of $d_{z^2} \to d_{x^2-y^2}$ transition	R_L^5	T^5
[Cu(dien) (HCO$_2$)]HCO$_2$	11.5	2.16	0.94
Cu D.M.P. Cl$_2$H$_2$O	13.1	2.23	0.89
[Cu(trien) (SCN)]NCS	14.1	2.61[a])	0.85[a])
[Cu(NH$_3$)$_4$H$_2$O]SO$_4$	16.0	2.34	0.86
[Cu(1,3 pn)H$_2$O]SO$_4$	16.0	2.57	0.79

[a]) Corrected[12] for the presence of a sulphur ligand.

F. π-Bonding

There is no X-ray crystallographic evidence (2) for a strictly square-coplanar CuL$_4$ chromophore (when L represents a purely sigma bonding ligand such as ammonia or ethylenediamine (but see ref. 29)) which suggest that in such CuL$_4$ chromophores some out-of-the-plane bonding potential still remains to be satisfied. Consequently, the square-coplanar stereochemistry will only exist with ligands capable of π-bonding to the copper(II) ion such as the oxygen atoms in acetonylacetone and the oxygen and nitrogen atoms in salicylaldimine which form the square coplanar complexes (2) Cu(acac)$_2$ and Cu(salim)$_2$. In complexes with ligands capable of π-bonding the four ligands may satisfy the additional bonding potential of the CuL$_4$ chromophore by out-of-the-plane π-bonding. The presence of π-bonding may be reflected in a number of ways; it can affect the length of the copper ligand bond; it will raise the energy of the d-orbitals involved in π-bonding (the d_{xz} and d_{yz} for out-of-the-plane and the d_{xy} orbital for in-plane bonding) relative to the pure σ-bonding sequence (Fig. 8), and it will lower, the orbital reduction factors determined from the ESR spectra as these may be related to the appropriate molecular orbital coefficients (2), (Table 6) (for pure σ-bonding $r_\parallel \approx r_\perp \approx 0.77$; for in-plane π-bonding $r_\parallel < r_\perp$ and for out-of-the-plane π-bonding $r_\perp < r_\parallel$). The short copper-oxygen bond-lengths of ca. 1.90 Å in Cu(3-Meacac)$_2$ (23) and CaCuSi$_4$O$_{10}$ (30) both of which involve a square-coplanar stereochemistry, are considered to arise both from the high tetragonal distortion present $T = <0.6$ and to the presence of significant out-of-the-plane π-bonding. In both complexes $r_\perp < r_\parallel$ and the mean energies $(26, 31)$ of the d_{xz}, $d_{yz} \rightarrow d_{x^2-y^2}$ transitions are reduced by 2.0—4.0 kK relative to the pure σ-bonding situation.

Table 6. *A, the g-value expressions for axial copper(II) complexes; B, M.O. expressions; C, the relationship between r-values and M.O. coefficients*

A. $\quad g_\perp = 2 - \dfrac{2\,r_\perp{}^2\lambda}{E_{d_{xz}, d_{yz} \to d_{x^2-y^2}}}$

$\quad\quad g_\parallel = 2 - \dfrac{8\,r_\parallel{}^2\lambda}{E_{d_{xy} \to d_{x^2-y^2}}}$

B. $\quad \psi_1 = \alpha\, d_{x^2-y^2} - \alpha^1\{ -\sigma_{x1} + \sigma_{y1} + \sigma_{x2} - \sigma_{y4} \}$
$\quad\quad$ etc.

C. $\quad r_\perp = \alpha.\beta \quad\quad\quad \alpha = \sigma\text{-bonding}$

$\quad\quad r_\parallel = \alpha.\beta_1 \quad\quad\quad \beta_1 = \text{in-plane } \pi\text{-bonding}$

$\quad\quad\quad\quad\quad\quad\quad\quad\quad \beta = \text{out-of-plane } \pi\text{-bonding}$

G. 1:1 and 1:2-Cu(acac)₂ Adducts

The ability of a square-coplanar chromophore, CuL_4, to take up a fifth or sixth ligand (32), B

$$CuL_4 + B \underset{K_1}{\overset{\Delta H_1}{\rightleftharpoons}} CuL_4B + B \underset{K_2}{\overset{\Delta H_2}{\rightleftharpoons}} CuL_4B_2$$

must be reflected in equilibrium constants and thermodynamic data. Such data does exist (33) for the bis(hexafluoroacetoxylacetonate) copper(II)-pyridine system and indicates clear evidence of 1:1 and 2:1 adduct formation. Unfortunately it is not possible to separate out the contribution that axial coordination makes to the overall addition of pyridine as the structure and bonding in the CuL_4 chromophore will also change on adduct formation. Added to which there is some uncertainty about the possible structure of these adducts in solution (Fig. 14 I—IV), although structures I and IV are believed (24) to dominate.

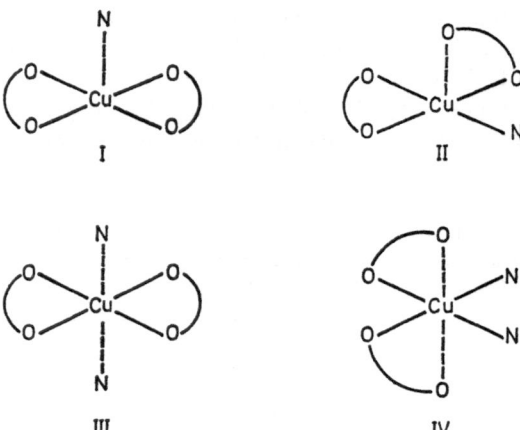

Fig. 14. The possible stereochemistries of the Cu(hfacac)₂py [I and II] and Cu (hfacac)₂py₂ [III and IV] complexes

Table 7. *Thermodymanic data for adduct formation between pyridine and Cu(hfaca)₂; 1:1 and 2:1 adducts*

	$-\Delta H$(kcal. mole^{-1})	K, M^{-1}
Cu(hfacac)₂ + py	11.8 ± 0.4	very large
Cu(hfacac)₂py + py	12.1 ± 0.3	570 ± 150

H. Off-the-z-Axis Bonding

The different types of complexes involving off-axis bonding are illustrated in Figs. 2 and 3. The two complexes (9) involving the dodecahedral structure of the CuO₄O₄ chromophore (Fig. 2c) have the structural and electronic (34) properties listed in Table 8. These are comparable, but do involve small significant differences. The shorter values of R$_S$ and longer values of R$_L$* in Cu(H₄A₄)ClO₄)₂ suggest a lower tetragonality or more nearly square-coplanar stereochemistry than in CaCu-(CH₃CO₂)₄6 H₂O a situation which is reflected in the lower energies of the d-d transitions. In neither complex is there any evidence from the r-values of out-of-the-plane π-bonding (consistent with the conformation of the in-plane oxygen ligands) but in both complexes the energy of the $d_{z^2} \rightarrow d_{x^2-y^2}$ transitions are surprisingly low (Table 8). These energy differences could be due to the shorter value of R$_S$ in Cu(H₄A₄) (ClO₄)₂

Table 8. *The structural and electronic properties of the complexes* $CaCuSi_4O_{10}$ *(square-coplanar)*, $CaCu(CH_3CO_2)_4\ 6\ H_2O$ *and* $Cu(H_4A_4)\ (ClO_4)_2(dodecahedral)$

Complex	R_S	R_L*)	T*)	$d_{z^2} \rightarrow$ $d_{xy}(d_{x^2-y^2})$
$CaCuSi_4O_{10}$	1.91	—	0.6	18.8
$Cu(H_4A_4)\ (ClO_4)_2$	1.93	2.88	0.67	16.0
$CaCu(CH_3CO_2)_4,\ 6\ H_2O$	1.97	2.79	0.71	12.2

Complex	$d_{xz}, d_{yz} \rightarrow$ $d_{xy}(d_{x^2-y^2})$	$d_{x^2-y^2}(d_{xy}) \rightarrow$ $d_{xy}(d_{x^2-y^2})$	$r_{\parallel}$	$r_{\perp}$
$CaCuSi_4O_{10}$	15.8	12.9	0.80	0.72
$Cu(H_4A_4)\ (ClO_4)_2$	17.5	13.6	0.79	0.77
$CaCu(CH_3CO_2)_4,\ 6\ H_2O$	14.4	10.8	0.78	0.78

*) Refers to off-the-z-axis ligands

but as this effect is small (Fig. 9) this is considered unlikely. The low values of the $d_{z^2} \rightarrow d_{xy}$ transitions, relative to that of 18.8 kK in the square coplanar $CaCuSi_4O_{10}$ (*26, 31*), suggests a higher tetragonality than square-coplanar and is considered to arise from sideways overlap (*34*) of a lone pair of electrons on the longbonded terminal oxygen atom of the carboxylate groups with the d_{z^2} orbital on the copper(II) ion (Fig. 15). With four terminal oxygen atoms bonding to the copper(II) ion, sufficient overlap is produced to lower the energy of the $d_{z^2} \rightarrow d_{xy}$ transition from that associated with a strictly square-coplanar (*26, 31*) chromophore, CuO_4, namely ca. 19.0 kK. A similar type of overlap occurs in $Cu(\alpha\text{-pic})_2$ $(ClCH_2CO_2)_2$ (*35*) and $Cu(\alpha\text{-pic})_2$ $(NO_3)_2$ (*36*), sufficient to lower the energy of the $d_{z^2} \rightarrow d_{x^2-y^2}$ transitions, 2.0 and 4.5 kK respectively, from that associated with a rhombic coplanar CuO_2N_2 chromophore (*37*), namely ca. 20.0 kK; and sufficient to justify a sixcoordinate description of the stereochemistry of $4+0+2*$ type in both complexes. The coordination in $Cu(\alpha\text{-pic})_2$ $(ClCH_2CO_2)_2$ $(4+0+2*)$ is then related to the $4+2+0$ type present in an elongated tetragonal octahedral stereochemistry and that in $Cu(\alpha\text{-pic})_2$ $(NO_3)_2$ $(4+0+2*)$ to a $4+1+0$ type present in a squarepyramidal stereochemistry. In the type of off-the-z-axis coordination present in the complexes of Fig. 3, i.e. in [Cudien (HCO_2)] HCO_2 (*10*) $(4+1+1*)$ and in $Cu(NO_3)_2\ 3\ H_2O$, $(4+1+2*)$ the effect of the off-axis coordination is masked by the four in-plane ligands plus that of the fifth ligand, but their effects should not be ignored, and justify a six and seven coordinate description of their stereochemistries, respectively.

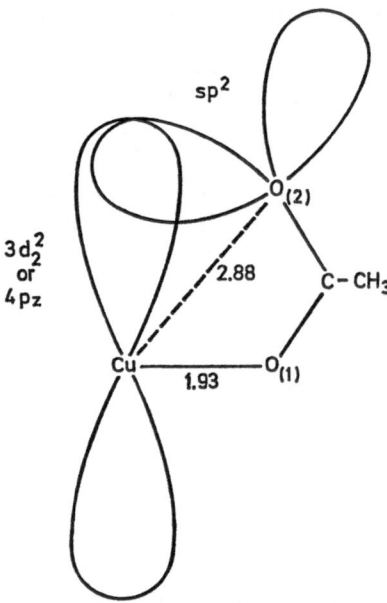

Fig. 15. The possible overlap of the 0(2) sp^2 hybridised orbital and the copper(II) 3 d_{z^2} and 4 p orbitals, along the z-axis

I. Overlap Criteria

There is some justification for considering axial ligands as bonding in in the evaluation (38) of the overlap integral $S(np_\sigma, nX_\sigma)$ where p_σ refers to a pure p-orbital on the ligand and X_σ refers to the sigma bonding orbitals on the copper(II) ion; these are shown in Fig. 16 for the 3d, 4s and 4p copper orbitals. The overlap at a distance R_L is 47% of that at R_s for 3d orbitals, 75% for 4s and nearly 100% for 4p, the latter arising as the overlap increases to a maximum above the value of R_S. Consequently, although bonding of these axial ligands decreases in the sequence $4p < 4s < 3d$, the relatively high value of the 4p overlap integral, suggests that these axial ligands must be considered as quite strongly bound. A similar consideration applies to off-the-z-axis ligands as long as the distance from the z-axis is not large, and $R_L^* \approx R_L$.

Consequently, the overlap of a lone-pair of electrons (sp^2 hybridised) will be greater with the 4p-orbital rather than with the d_{z^2}-orbital, as suggested (35) in Fig. 15.

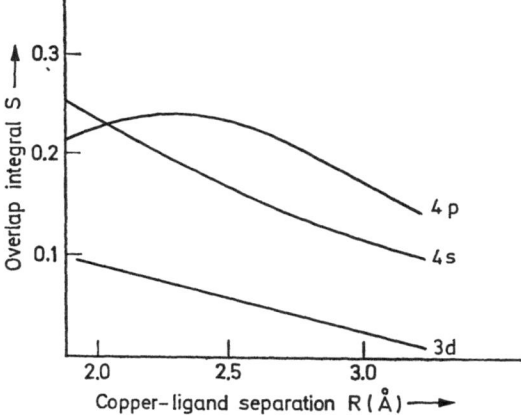

Fig. 16. The diatomic overlap integrals $S(np_\sigma 3d_\sigma\text{-}4s\text{-}4p)$ between a copper(II) ion and ligand atoms (nitrogen ligands only)

J. — *Prolate Ellipsoidal Shape of the Copper(II) Ion.* The above data suggest that the presence of out-of-the-plane ligands in a wide variety of axial copper(II) complexes has an appreciable influence on the electronic properties of the copper(II) ion and despite the greater bond-lengths involved represent an appreciable bonding effect measured in terms of overlap. A rationale of these properties can be sought in terms of the lack of spherical symmetry (*2, 14, 39*) of the d^9 electron configuration of the copper(II) ion. If the copper (II)ion in an elongated tetragonal octahedral stereochemistry is considered to have a prolate ellipsoidal shape (Fig. 17), then the observed copper-ligand bond-lengths

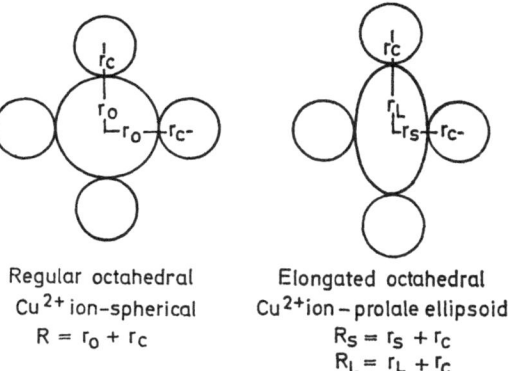

Fig. 17. The copper and ligand atom covalent radii in octahedral and elongated octahedral stereochemistries

B. J. Hathaway

will reflect the assymetry of the effective radius of the ion in the direction of the relevant bonds. This difference is well illustrated in the bond-length data for the complexes of Ni(NH$_3$)$_4$ (NO$_2$)$_2$ (40) and Cu(NH$_3$)$_4$ (NO$_2$)$_2$ (41) (Fig. 18) as both complexes involve a MN$_4$N$_2$-chromophore.

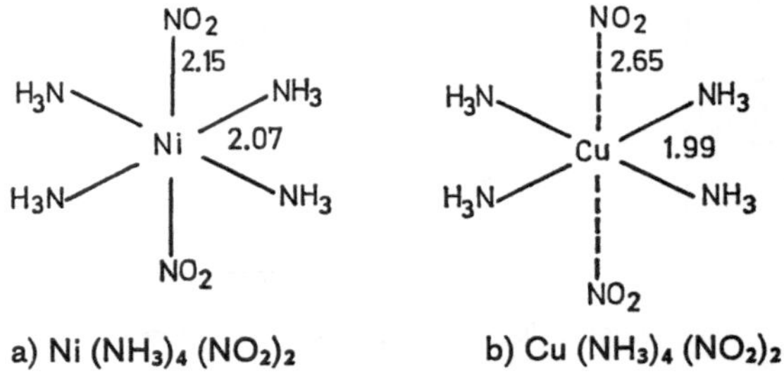

Fig. 18. The molecular structures of (a) Ni(NH$_3$)$_4$ (NO$_2$)$_2$ and (b) Cu(NH$_3$)$_4$(NO$_2$)$_2$

As the nickel(II) ion has a spherically symmetrical d^8 electron configuration, the differences in the nickel-nitrogen bond lengths ($R_L - R_s \approx$ 0.1 Å simply reflect the presence of non-equivalent ligands. In the copper(II) complex the copper-nitrogen bond lengths involve a much greater bond-length difference ($R_L - R_s = 0.66$ Å) and reflects the non-spherical symmetry of the copper(II) ion.

K. — *Conclusion.* The prolate ellipsoidal description of the shape of the copper (II) ion accounts for the formation of four short in-plane copper ligand bonds, but the additional out-of-the-plane bonding potential of the CuL$_4$ chromophore may then be satisfied in a number of ways, viz.: by coordination of a single fifth ligand (square pyramidal), by semi-coordination of a fifth and sixth ligand (elongated tetragonal octahedral) by out-of-plane π-bonding (square coplanar) or by off-the-z-axis coordination of from one to four additional ligands to give the range of coordination numbers from five to eight.

References

1. *Hatfield, W. E., Whyman, R.*: Trans. Met. Chem. **5**, 47 (1969).
2. *Hathaway, B. J., Billing, D. E.*: Coord. Chem. Rev. **5**, 143 (1970).
3. *Cotton, F. A., Wilkinson, G.*: Advan. Inorg. Chem., 3rd Ed. Interscience, London 1972.
4. *MacKay, K. M., MacKay, R. A.*: Introduction Mod. Inorg. Chem. 2nd Ed. London: Intertext Books 1972.

5. *Dunitz, J. D., Orgel, L. E.:* Nature *179*, 462 (1957).
6. *Davey, G., Stephens, F. S.:* J. Chem. Soc. (A) 1917 (1971).
7. *Stepehens, F. S.:* J. Chem. Soc. (A) 2081 (1969).
8. *Cameron, A. F., Taylor, D. W., Nuttall, R. H.:* J. C. S., Dalton 58 (1972).
9. *Langs, D. A., Hare, C. B.:* Chem. Comm. 890 (1967); *Österberg, R., Björberg, B., Söderquist, R.:* Chem. Comm. 1410 (1970).
10. *Davey, G., Stephens, F. S.:* J. Chem. Soc (A) 103 (1971).
11. *Morosin, B.:* Acta Cryst. B 26, 1203 (1970).
12. *Procter, I. M., Hathaway, B. J., Nicholls, P.:* J. Chem. Soc (A) 1768 (1968).
13. *Pajunen, A.:* Suomen Kem. 40, 32 (1967).
14. *Hathaway, B. J., Tomlinson, A. A. G.:* Coord. Chem. Rev. 5, 1 (1970).
15. *—, Underhill, A. E.:* J. Chem. Soc. 3091 (1961).
16. *Procter, I. M., Hathaway, B. J., Hodgson, P. G.:* J. Inorg. Nucl. Chem., a in the press 1973.
17. *Hathaway, B. J., Webster, D. E.:* Proc. Chem. Soc. 14 (1963).
18. *Brown, D. S., Lee, J. D., Melson, B. G. A.:* Acta Cryst. B 24, 730 (1968).
19. *Lee, J. D., Brown, D. S., Melson, B. G. A.:* Acta Cryst. B 25, 1595 (1969).
20. *Hathaway, B. J., Nicholls, P., Barnard, D.:* Spectrovision *No.* 22, 4 (1969).
21. *Fenton, D. E., Truter, M. R., Vickery, B. L.:* Chem. Comm. 93 (1971).
22. *Gillard, R. D., Rogers, D., Diamond, R. D., Williams, D. J.:* Acta Cryst. *A 67* (1963).
23. *Robertson, I., Truter, M. R.:* J. Chem. Soc. (A) 309 (1967).
24. *Dudley, R. J., Hathaway, B. J., Hodgson, P. G.:* unpublished work.
25. *Dudley, R. J., Hathaway, B. J.:* J. Chem. Soc. (A) 1725 (1970).
26. *Hathaway, B. J., Billing, D. E., Dudley, R. J.:* J. Chem. Soc (A) 1420 (1970).
27. *Dudley, R. J., Hathaway, B. J., Hodgson, P. G., Tomlinson, A. A. G.:* Autumn Meeting Chem. Soc., York 1971.
28. *Beattie, I. R., Gilson, T. R., Ozin, G.:* J. Chem. Soc. (A) 534 (1969); *Dunsmuir, J. T. R., Lane, A. P.:* J. Chem. Soc. (A) 2724 (1971); *Adams, D. M., Newton, D. C.:* J. Chem. Soc. (A) 3507 (1971).
29. *Hathaway, B. J., Stephens, F. S.:* J. Chem. Soc. (A) 884 (1970).
30. *Pabst, A.:* Acta Cryst. *12*, 733 (1959).
31. *Billing, D. E., Hathaway, B. J., Nicholls, P.:* J. Chem. Soc. (A) 316 (1969).
32. *Graddon, D. P.:* Coord. Chem. Rev. *4*, 1 (1969).
33. *Partenheimer, W., Drago, R. S.:* Inorg. Chem. *9*, 47 (1970).
34. *Billing, D. E., Hathaway, B. J., Nicholls, P.:* J. Chem. Soc (A) 1877 (1970); *Dudley, R. J., Hathaway, B. J., Hodgson, G.:* J. Chem. Soc. (A) 1355 (1971).
35. *Davey, G., Dudley, R. J., Hathaway, B. J.:* J. Chem. Soc. (A) 1446 (1971).
36. *Dudley, R. J., Fereday, R. J., Hathaway, B. J., Hodgson, P. G.:* J. C. S. Dalton, accepted for publication 2/1661 (1972).
37. — — — — J. C. S. Dalton, 1341 (1972).
38. *Smith, D. W.:* J. Chem. Soc. (A) 1499 (1970).
39. *Gillespie, R. J.:* J. Chem. Soc. 4672 (1963).
40. *Porai-Koshitz, M. A., Dikaneva, L. M.:* Kristallgrafiya *4*, 650 (1959).
41. *Bukovska, M., Porai-Koshitz, M. A.:* Zhurstrukt. Khim. *7*, 712 (1961).

Received April 26, 1972

Two Symmetry Parameterizations of the Angular-Overlap Model of the Ligand-Field

Relation to the Crystal-Field Model

C. E. Schäffer

Department I, (Inorganic Chemistry), The H. C. Ørsted Institute, University of Copenhagen, Universitetsparken 5, DK-2100 Copenhagen Ø, Denmark

Table of Contents

1. Introduction. The Relevance of Ligand-Field Models

The main reason why ligand-field models are important is the qualitative experimental fact that hundreds of compounds containing a transition metal ion with a partially filled d or f shell have ground states and lower excited states whose symmetries can be predicted on the basis of a first-order perturbation model (using a basis set of l functions, $l = 2$ or 3) once the geometries of the molecules are known (1—4).

Ligand-field models have also been successfully applied quantitatively in the sense that they have been used for parameterizing experimental results. This application, however, requires carefully chosen experiments and a certain skill with the computer in order to go directly from the experimental results to the best fitting parameters while at the same time evaluating the variances of these parameters on the basis of the uncertainties of the experiments.

For molecular systems of high symmetry, it is useful to apply group-theoretical considerations to establish the number of one-electron parameters required for parameterizing the relative energies of the electrons of the partially filled shell of a given system. It is assumed that this shell transforms under the molecular point group in the same way as does the d or f shell of the central ion [Ref. (5) p. 199]. For such systems these one-electron parameters required by symmetry make up the most general model parameterization possible for the energies of the partially filled shell. Apart from certain difficulties with interelectronic repulsion [Ref. (5) p. 229] within this shell, the parameters for a given compound can, at least in principle, be determined by experiment.

As the symmetry of the system becomes lower, the number of symmetry-determined parameters increases rapidly and it soon becomes impossible to determine them experimentally.

This is one of the situations which demands more specific ligand-field models. These specific models introduce linear dependencies (6) between the symmetry-determined parameters, hopefully in a meaningful way, thus reducing the number of independent parameters.

The models endeavour to introduce semiempirical parameters[1] which, in addition to having a chemical appeal or even significance, are transferable or at least comparable from one molecule to another.

Therefore the other situation in which specific models are important is that in which the symmetry is high enough to give a low number of parameters, but where many related molecules exist so that experimentally derived parameters can be compared (7). In this way the specific models, probably in contradistinction to more sophisticated molecular orbital treatments, may serve as a means of making classifications within inorganic chemistry (7, 8). This is the main scope of these models.

The *angular-overlap model* (AOM) is one such specific model whose parameters have been chosen so as to refer to the central ion-to-ligand bonding process (9). However, it must be remembered that AOM still is a first-order perturbation model (10) but, one may say, a zeroth-order molecular orbital model (8).

It is a characteristic feature of AOM that its matrix elements may be factorized into a parametric factor, which one may call the radial factor, and a factor which makes up the coefficient to this semiempirical parameter (8—11). The latter factor depends only on angular coordinates: The angular properties of the l orbitals, and the relative astronomical

---

[1] It would seem more appropriate to call the model semiempirical and the parameters empirical.

positions of the ligands and their rotational orientations with respect to the bond (10).

In the present paper AOM is reparameterized for the case when the central ion-to-ligand bonds have linear symmetry and a comparison with the conventional perturbation model is made. It is shown that an electrostatic contribution to the semiempirical parameters can never be evaluated from experiments determining energy parameters. Further, some of the problems associated with the zero points for the energy (12) are discussed [Ref. (13) p. 130].

2. The Angular-Overlap Model (AOM)

The total core field $U(x,y,z)$ of a molecular system (14) containing a central ion forming essentially heteropolar bonds to the surrounding ligands can be written [Ref. (8) p. 370] as

$$U(x,y,z) = U(r) + V(r) + A(x,y,z) \qquad (1)$$

where $U(r)$ and $V(r)$ are spherically symmetrical terms, $U(r)$ representing the core field of the central ion, $V(r)$ the essential part of the ligand field, and $A(x,y,z)$ the part of the ligand field which contains the non-spherically symmetrical terms. This part is called A because it is the part which AOM endeavours to account for.

The assumptions of the model may be summarized as follows (10):

I. $A(x,y,z)$ can be accounted for by a first-order perturbation, either upon a d basis or upon an f basis.

II. If the l basis is the usual real set of functions (8, §3) defined relative to a coordinate system XYZ, then the perturbation matrix of the ligand field caused by a ligand placed on the Z axis is diagonal.

III. Perturbation contributions from different ligands are additive.

If the l basis is quantized with respect to an axis, the Z axis, say, they fall into the classes $\sigma, \pi, \delta \ldots$, according as their λ values are $0, 1, 2, \ldots$. For $\lambda = 0$ there is only one function, but for $\lambda > 0$ there is a pair of functions for each λ value and, by fixing the direction of the X axis and thereby the origin of the azimuthal polar coordinate φ, these pairs can be completely specified by their φ dependence [$\sin(\lambda\varphi)$ and $\cos(\lambda\varphi)$] as sine and cosine functions. We shall use these standard functions (8 §3) with a standard (10) order $\sigma, \pi s, \pi c, \delta s, \delta c, \ldots$ and a standard numbering[2] $0, 1, 2, 3, 4, \ldots$, so that even numbers (2λ) refer to λ cosine

---

[2] We note that the present numbering, but not the present order, is different from that used previously [Ref. (8) p. 371].

functions (λc), and odd numbers ($2\,\lambda - 1$) refer to λ sine functions (λs). σ functions ($\lambda = 0$) thereby fall within the cosine class and we obtain a kind of parity rule, a plus-minus rule, for the products of such functions [Ref. (6) p. 285].

$$c \times c = s \times s = c \times \sigma = c, \quad s \times c = s \times \sigma = s \tag{2}$$

where in Eq. (2) the terms should be represented by the parity of the numbering symbol.

Assumptions I and II of the AOM may be expressed by the diagonal energy matrix

$A^{Z'}$	$\|0'>$	$\|1'>$	$\|2'>$	$\|3'>$	$\|4'>$	$\cdot$	$\cdot$
$<0'\|$	$e(r) + e_{0'}$						
$<1'\|$		$e(r) + e_{1'}$					
$<2'\|$			$e(r) + e_{2'}$				
$<3'\|$				$e(r) + e_{3'}$			
$<4'\|$					$e(r) + e_{4'}$		
$\cdot$						$\cdot$	
$\cdot$							$\cdot$

$$\tag{3}$$

expressing the perturbation energies of a real l basis (defined with respect to the primed coordinate system X′Y′Z′) caused by a ligand upon the Z′ axis. In Eq. (3) $e(r)$ represents the spherical term $V(r)$ of Eq. (1), but since the $e_{t'}$ parameters do not add up to zero, they also include a spherical term. We may write $A(x, y, z)$ of Eq. (1) as

$$A(x, y, z) = A(r) + \bar{A}(x, y, z) \tag{4}$$

but in the present section we treat $A(x, y, z)$ as an entirety and postpone further consideration of Eq. (4) till Sect. 3.

When the symmetry about the central ion-to-ligand bond is linear, we say that the ligand is linearly ligating (13). In this case one aspect of the symmetry basis for the AOM becomes apparent. Atomic orbitals of the central ion — and those of the ligand as well as those molecular orbitals which arise by their interaction — may now be classified according to irreducible representations of the group $C_{\infty v}$. This group is not a true point group (15) since its elements of symmetry do not intersect at a point, but rather in the line which makes up the axis of symmetry. Therefore the atomic orbitals with the two different centers and the molecular orbitals are all associated with one of the good quantum numbers $\lambda = 0$, 1, 2, and for f electrons 3, and we can speak of a σ interaction, a π interaction, and so on, between the central ion and the ligand orbitals.

The other aspect of the symmetry basis for AOM, and we now return to the more general case of Eq. (3) where the ligands are not necessarily linearly ligating, is that expressed by assumption III (p. 71). This assumption makes AOM an additivity model or a superposition model (16). This was discussed thoroughly in a previous paper (10) where it was shown that the general matrix element of AOM within an l basis contains the rotation[3]) matrices $D^{(l)}[R(\varphi, \theta, \psi)]$ or, in other words, the matrices of the $(2l+1)$-dimensional irreducible representation of the three-dimensional rotation group. We shall not pursue this general case here.

However, in order to take into account the perturbation effect due to more ligands, certain rotations of the l basis are necessary (8, 10). We consider the XYZ coordinate system as space-fixed, and we introduce a new coordinate system X'Y'Z' with which we associate a primed l

Table 1. *Angular overlap matrix for p orbitals. — The columns give the direction cosines (γ, β, α) of the primed axes Z', Y', X', relative to the unprimed axes Z, Y, X and the rows give the reverse relationship. — The columns also give the coefficients of decomposition of the primed functions into linear combinations of unprimed ones and again the rows give the reverse relationship. $\delta = +\sqrt{\alpha^2 + \beta^2}$. — The direction cosines (γ, β, α) are equal to the Cartesian coordinates (z, y, x) (measured in the unprimed coordinate system) at the point at which the Z' axis cuts the sphere $r^2 = x^2 + y^2 + z^2 = 1$. — For a description of the unprimed and primed coordinate systems, see also p. 74*

$F^{(p)}$ (γ, β, α)			$\|(z')>$ $\|(\sigma)'>$ $\|0'>$	$\|(y')>$ $\|(\pi s)'>$ $\|1'>$	$\|(x')>$ $\|(\pi c)'>$ $\|2'>$
$<(z)\|$	$<(\sigma)\|$	$<0\|$	γ	0	$-\delta$
$<(y)\|$	$<(\pi s)\|$	$<1\|$	β	$\dfrac{\alpha}{\delta}$	$\dfrac{\beta\gamma}{\delta}$
$<(x)\|$	$<(\pi c)\|$	$<2\|$	α	$-\dfrac{\beta}{\delta}$	$\dfrac{\alpha\gamma}{\delta}$

[3]) Throughout this paper semi-heavy type will be used for matrices and matrix elements, gothic type for irreducible tensor operators (6, 15) and Gill sans for the ligand field operator, for projection operators, and for rotation operators (11).

basis $0'$, $1'$, $2'$, $3'$, $4'$, ... The axes of the primed coordinate system $Z'X'Y'$ are parallel (8) to the respective infinitesimal direction vectors of the right-handed polar coordinate system $r\,\theta\,\varphi$ associated with the Cartesian XYZ system, and the Z' axis has the polar coordinates (θ, φ) or the direction cosines $(\alpha, \beta, \gamma) = (\sin\theta\,\cos\varphi,\ \sin\theta\,\sin\varphi,\ \cos\theta)$ referred to the XYZ system. This means that the X' axis lies in the ZZ' plane — pointing downwards towards increasing θ values — and that the Y' axis is perpendicular to the Z axis (8, 9).

There exists a linear relationship (8) between the primed and the unprimed real l basis which can be expressed by the matrix equations[4]

$$\boldsymbol{f'} = \boldsymbol{f}\ \boldsymbol{F}$$
$$\boldsymbol{f} = \boldsymbol{f'}\ \tilde{\boldsymbol{F}} \tag{5}$$

where $\boldsymbol{f'}$ and $\boldsymbol{f}$ are row matrices containing the functions in standard order and $\boldsymbol{F}$ is an orthogonal matrix, called the angular-overlap matrix (8, 9) for reasons not to be discussed here. $\boldsymbol{F}$ is a function of (θ, φ) or (α, β, γ). The $\boldsymbol{F}^{(p)}$ (γ, β, α) and $\boldsymbol{F}^{(d)}$ (γ, β, α) matrices referring to p and d orbitals are given in Tables 1 and 2.

Eq. (5) may be written out as follows

$$u = \sum_{t'=0}^{t'=2l} t'\ \tilde{F}_{t'u} = \sum_{t'=0}^{t'=2l} t'\ F_{ut'} \tag{6}$$

We now consider the perturbation A^k upon the unprimed basis set (8—10), from the ligand $L(k)$ upon the Z' axis, whose direction cosines are $(\alpha_k, \beta_k, \gamma_k)$. These direction cosines are also equal to the Cartesian coordinates (x_k, y_k, z_k) of the ligand when this is projected upon the unit sphere along the direction toward the origin. The energy matrix with respect to the particular primed basis set is the diagonal matrix of Eq. (3). For the unprimed basis set we have, using Eq. (6),

$$(u|A^k|v) = \sum_{t'=0}^{t'=2l} F_{ut'}^k\ F_{vt'}^k\ <t'|A^k|t'> = \delta_{uv}\ e(r) + \sum_{t=0}^{t=2l} F_{ut}^k\ F_{vt}^k\ e_t \tag{7}$$

---

[4] In Eq. (5) we use the matrix notation of Refs. (8) and (10). An alternative way of writing Eqs. (5) is
$$|f'\} = |f\}\ \{f|f'\}$$
$$|f\} = |f'\}\ \{f'|f\}$$
using the set notation of Ref. (6) p. 261, and thereby showing that $\boldsymbol{F}$ is the matrix of the overlap integrals between the unprimed and the primed real basis sets.

Table 2. Angular overlap matrix for d orbitals. Interrelationship between unprimed and primed d functions, where the interrelations between the corresponding coordinate systems have been given in Table 1. $\delta = +\sqrt{\alpha^2 + \beta^2}$. See also text for Table 1

$F^{(d)}(\gamma, \beta, \alpha)$	$\begin{array}{c}\|(z^2)'\rangle\\ \|(\sigma)'\rangle\\ \|0'\rangle\end{array}$	$\begin{array}{c}\|(yz)'\rangle\\ \|(\pi s)'\rangle\\ \|1'\rangle\end{array}$	$\begin{array}{c}\|(zx)'\rangle\\ \|(\pi c)'\rangle\\ \|2'\rangle\end{array}$	$\begin{array}{c}\|(xy)'\rangle\\ \|(\delta s)'\rangle\\ \|3'\rangle\end{array}$	$\begin{array}{c}\|(x^2-y^2)'\rangle\\ \|(\delta c)'\rangle\\ \|4'\rangle\end{array}$
$\begin{array}{c}\langle(z^2)\|\\ \langle(\sigma)\|\\ \langle 0\|\end{array}$	$(1/2)(2\gamma^2-\delta^2)$	0	$-\sqrt{3}\gamma\delta$	0	$(\sqrt{3}/2)(\delta^2)$
$\begin{array}{c}\langle(yz)\|\\ \langle(\pi s)\|\\ \langle 1\|\end{array}$	$\sqrt{3}\,\beta\gamma$	$\dfrac{\gamma\alpha}{\delta}$	$\dfrac{\beta(\gamma^2-\delta^2)}{\delta}$	$-\alpha$	$-\beta\gamma$
$\begin{array}{c}\langle(zx)\|\\ \langle(\pi c)\|\\ \langle 2\|\end{array}$	$\sqrt{3}\,\gamma\alpha$	$-\dfrac{\beta\gamma}{\delta}$	$\dfrac{\alpha(\gamma^2-\delta^2)}{\delta}$	β	$-\gamma\alpha$
$\begin{array}{c}\langle(xy)\|\\ \langle(\delta s)\|\\ \langle 3\|\end{array}$	$\sqrt{3}\,\alpha\beta$	$\dfrac{\alpha^2-\beta^2}{\delta}$	$\dfrac{2\alpha\beta\gamma}{\delta}$	$\dfrac{(\alpha^2-\beta^2)\gamma}{\delta^2}$	$\dfrac{\alpha\beta(2\gamma^2+\delta^2)}{\delta^2}$
$\begin{array}{c}\langle(x^2-y^2)\|\\ \langle(\delta c)\|\\ \langle 4\|\end{array}$	$(\sqrt{3}/2)(\alpha^2-\beta^2)$	$-\dfrac{2\alpha\beta}{\delta}$	$\dfrac{(\alpha^2-\beta^2)\gamma}{\delta}$	$\dfrac{-2\alpha\beta\gamma}{\delta^2}$	$\dfrac{(\alpha^2-\beta^2)(2\gamma^2+\delta^2)}{2\delta^2}$

where the double summation which arises through the transformation of u and v into the primed basis reduces to a single summation because of the diagonality of A^k with respect to the primed basis. F^k of Eq. (7) is a short way of writing $F(\gamma_k, \beta_k, \alpha_k)$. The expression of the last line of Eq. (7) is obtained by using the orthonormality of the rows of the F matrix. It shows that the zero point does not influence the non-diagonal elements. Further (10), in Eq. (7) $F_{ut'} = F_{ut}$ and the prime has been removed from the dummy index.

We are now able to take into account the perturbation effect of more ligands upon our standard basis, the unprimed basis. This is done by introducing assumption III which states that we have only to sum the expressions of Eq. (7) over k. It is important to observe that each new ligand corresponds to a new position of the Z' axis and thereby to a new primed basis set, but this basis set has been eliminated from the formulae, which only contain the direction cosines (α_k, β_k, γ_k) of each Z' axis and thereby are fully characterized by the astronomical (directional) coordinates of each ligand. The final formula for equal ligands is

$$(u|A|v) = \sum_k (u|A^k|v) = \delta_{uv} N e(r) + \sum_{k=1}^{k=N} \sum_{t=0}^{t=2l} F_{ut}^k F_{vt}^k e_t \qquad (8)$$

where N is the number of ligands.

This formula may be made valid also for different ligands by adding an index on the $e(r)$ and an extra index on the e_t parameters and splitting up N into the different types of ligand which occur.

3. Average Energies, Zero Points of Energy and Linearly Ligating Ligands

It is useful, particularly for the comparisons to come, to express the matrix elements of Eq. (8) relative to the average energy of the l shell.

Following Eq. (4) we put a bar over the angular-overlap operator to indicate that this new zero point of energy has been used

$$(u|\bar{A}|v) = (u|A|v) - [\delta_{uv}/(2l + 1)] \sum_{t=0}^{t=2l} (i|A|i)$$

$$= \sum_k \{(u|A^k|v) - [\delta_{uv}/(2l + 1)] \sum_{t=0}^{t=2l} (i|A^k|i)\} \qquad (9)$$

$$= \sum_k (u|\bar{A}^k|v)$$

and we note that the consequence of the new zero point for A is that a particular perturbation contribution from a given ligand, i.e. a given k, becomes measured relative to the perturbation contribution from that ligand averaged over the l orbitals. Further, since the perturbation is expressed as a sum of terms of σ type, πs type, πc type, and so on, each such term may be considered as having its own zero point.

Written in full, the expression for the matrix element of $\bar{A}^k$ is

$$(u|\bar{A}^k|v) = \sum_{t=0}^{t=2l} \{F_{ut}^k F_{vt}^k - \delta_{uv}[1/(2l+1)]\} e_t$$

$$\equiv \sum_{t=0}^{t=2l} \overline{F_{ut}^k F_{vt}^k} e_t \tag{7}$$

where the symbol $\equiv$ signifies that we have an equality by definition and where the unbarred and barred coefficients to e_t are equal for non-diagonal matrix elements of $\bar{A}^k$. From the orthonormality of the rows of the F matrix $(8, 9)$ it follows that

$$\sum_{t=0}^{t=2l} F_{ut}^k F_{vt}^k = \delta_{uv} \tag{10a}$$

and that

$$\sum_{t=0}^{t=2l} \overline{F_{ut}^k F_{vt}^k} = 0 \tag{10b}$$

also for $u = v$, which is another way of stating that the energy differences between the $(2l+1)$ orbitals only require $2l$ independent parameters. We note that from Eq. (10b) it follows that some of the coefficients $\overline{(F_{ut}^k)^2}$ must be negative.

The final formula, corresponding to Eq. (8) but with the energies measured relative to the average energy of the l shell, is obtained from Eq. (7) by summing over the N ligands.

$$<u|\bar{A}|v> = \sum_{k=1}^{k=N} \sum_{t=0}^{t=2l} \overline{F_{ut}^k F_{vt}^k} e_t \tag{8}$$

$$= \sum_{t=0}^{t=2l} \{- \delta_{uv}[N/(2l+1)] + \sum_{k=1}^{k=N} F_{ut}^k F_{vt}^k\} e_t$$

In Eqs. (7) and (8) the spherical part of the perturbation has been completely eliminated [cf. the comments accompanying Eq. (3)].

As an example we write the diagonal element of a pu orbital for which the expression, using Eqs. $(\overline{7})$ and $(10\,\mathrm{b})$, is

$$
\begin{aligned}
<pu|\bar{A}^k|pu> &= \overline{(F_{u0}^k)^2}\, e_0 + \overline{(F_{u1}^k)^2}\, e_1 + \overline{(F_{u2}^k)^2}\, e_2 \\
&= \overline{(F_{u0})^2}\, e_0 + \overline{(F_{u1}^k)^2}\, e_1 - [\overline{(F_{u0}^k)^2} + \overline{(F_{u1}^k)^2}]\, e_2 \quad (11) \\
&= \overline{(F_{u0}^k)^2}\, (e_0 - e_2) + \overline{(F_{u1}^k)^2}\, (e_1 - e_2)
\end{aligned}
$$

so that the e_2 or $e_{\pi c}$ parameter has been used as the zero point for the others.

It is of importance for the comparison with other models to see what formally happens if one ignores more parameters. We illustrate this by looking at a water molecule with the oxygen on the Z' axis and the molecular plane coinciding with $Z'X'$ plane[5]. This water molecule perturbs a d shell with the available part of its outer shell of electrons. In a book-keeping sense, two of the electron pairs around the oxygen are engaged in bonding to the hydrogens and the remaining two give rise to a σ and a πs perturbation of the d basis (10). Using Eq. $(\overline{7})$, the diagonal element for the du orbital can be written

$$
\begin{aligned}
<du|\bar{A}^k|du> &= \sum_{t=0}^{t=4} \overline{(F_{ut}^k)^2}\, e_t = \overline{(F_{u0}^k)^2}\, e_0 + \overline{(F_{u1}^k)^2}\, e_1 \\
&- \frac{\overline{(F_{u0}^k)^2} + \overline{(F_{u1}^k)^2}}{\overline{(F_{u2}^k)^2} + \overline{(F_{u3}^k)^2} + \overline{(F_{u4}^k)^2}} \left[\overline{(F_{u2}^k)^2}\, e_2 + \overline{(F_{u3}^k)^2}\, e_3 + \overline{(F_{u4}^k)^2}\, e_4 \right] \\
&= (F_{u0}^k)^2 \left[e_0 - \frac{\overline{(F_{u2}^k)^2}\, e_2 + \overline{(F_{u3}^k)^2}\, e_3 + \overline{(F_{u4}^k)^2}\, e_4}{\overline{(F_{u2}^k)^2} + \overline{(F_{u3}^k)^2} + \overline{(F_{u4}^k)^2}} \right] \quad (12) \\
&+ (F_{u1}^k)^2 \left[e_1 - \frac{\overline{(F_{u2}^k)^2}\, e_2 + \overline{(F_{u3}^k)^2}\, e_3 + \overline{(F_{u4}^k)^2}\, e_4}{\overline{(F_{u2}^k)^2} + \overline{(F_{u3}^k)^2} + \overline{(F_{u4}^k)^2}} \right]
\end{aligned}
$$

It is seen that deleting some of the parameters (in this case e_2, e_3, and e_4) means that the remaining ones (in this case e_0 and e_1) become measured relative to a weighted mean of the deleted ones, the weighting being proportional to the square of the angular-overlap integral with the particular d orbital, in this case u. This dependence on a particular d orbital is, of course, an impossible situation unless the parameters e_2, e_3, and e_4 are for practical purposes negligible compared with e_0 and e_1. However, if the parameters of the angular-overlap model can be taken seriously, e_2, e_3, and e_4 should vanish in our present example.

---

[5] One may say, by analogy with the expression, linearly ligating, that this water molecule is planarly ligating. There is another way in which water molecules have been found to bind (17). This way may, in a self-explanatory expression, be called tetrahedrally ligating.

When the symmetry about the central ion-to-ligand bond is linear (linear ligation), and it is a symmetry property of the system that the energy depends only upon λ so that for $\lambda > 0$ degeneracies occur, then we define for d electrons

$$\overline{F^k_{u1} F^k_{v1}} + \overline{F^k_{u2} F^k_{v2}} = F^k_{u1} F^k_{v1} + F^k_{u2} F^k_{v2} - \delta_{uv}(2/2l + 1)$$

$$\equiv F^k_{u\pi} F^k_{v\pi} - \delta_{uv}(2/2l + 1) \equiv \overline{F^k_{u\pi} F^k_{v\pi}};$$

$$\overline{F^k_{u3} F^k_{v3}} + \overline{F^k_{u4} F^k_{v4}} = F^k_{u3} F^k_{v3} + F^k_{u4} F^k_{v4} - \delta_{uv}(2/2l + 1) \qquad (13)$$

$$\equiv F^k_{u\delta} F^k_{v\delta} - \delta_{uv}(2/2l + 1) \equiv \overline{F^k_{u\delta} F^k_{v\delta}};$$

since in this case by symmetry

$$e_1 = e_2 \equiv e_\pi \; ; \qquad e_3 = e_4 \equiv e_\delta \qquad (14)$$

For a matrix element of A^k we then have

$$<du|A^k|dv> = F^k_{u\sigma} F^k_{v\sigma} e_\sigma + F^k_{u\pi} F^k_{v\pi} e_\pi + F^k_{u\delta} F^k_{v\delta} e_\delta \qquad (15)$$

and, for a matrix element of $\bar{A}^k$, the barred equation

$$<du|\bar{A}^k|dv> = \overline{F^k_{u\sigma} F^k_{v\sigma}} e_\sigma + \overline{F^k_{u\pi} F^k_{v\pi}} e_\pi + \overline{F^k_{u\delta} F^k_{v\delta}} e_\delta \, . \qquad (\overline{15})$$

Alternatively, using Eqs. (10), (13), and (14),

$$<du|\bar{A}^k|dv> = \overline{F^k_{u\sigma} F^k_{v\sigma}} (e_\sigma - e_\delta) + \overline{F^k_{u\pi} F^k_{v\pi}} (e_\pi - e_\delta) \qquad (\overline{16}a)$$

$$\equiv \overline{F^k_{u\sigma} F^k_{v\sigma}} e'_\sigma + \overline{F^k_{u\pi} F^k_{v\pi}} e'_\pi$$

or

$$<du|\bar{A}|dv> = \sum_{k=1}^{k=N} <du|\bar{A}^k|dv> = \sum_{k=1}^{k=N} \sum_{t=\sigma,\pi} \overline{F^k_{ut} F^k_{vt}} e'_t \qquad (\overline{16}b)$$

valid also for $u = v$ [cf. Eqs. ($\overline{7}$) and (10)]. Eq. (15) corresponds to Eq. (7), Eqs. ($\overline{15}$) and ($\overline{16}a$) to Eq. ($\overline{7}$), and Eq. ($\overline{16}b$) to Eq. ($\overline{8}$). There are no equations with the unbarred operator corresponding to Eqs. ($\overline{16}a$) and ($\overline{16}b$), but this is of no applicational consequence since the unbarred operator may equally well be used for energy differences and non-diagonal elements, as shown below [Eqs. (21) and (22)]. The Eqs. (15), ($\overline{15}$), ($\overline{16}a$) and ($\overline{16}b$) express the matrix elements in terms of the *angular overlap model symmetry parameters* for *linearly ligating ligands* [see also

p. 85]. As we shall see, the *angular-overlap model* for *linearly ligating ligands* is in a parametrical sense equivalent to the point-charge or point-dipole electrostatic model. For d electrons the coefficients of Eqs. $(\overline{15})$ and $(\overline{16a})$, applying to a ligand with a position (α, β, γ), have been derived by the use of Table 2 and Eqs. $(\overline{7})$, (13), (14) and $(\overline{16a})$; they are given in Tables 3—5.

It may be noted that Eqs. $(\overline{16a})$ and $(\overline{16b})$ can be obtained from $(\overline{7})$ and $(\overline{8})$, respectively, by the substitution

$$e_0 \to e'_\sigma; \quad e_1 = e_2 \to e'_\pi; \quad e_3 = e_4 \to 0. \tag{17}$$

When $\overline{A}^k$ of Eqs. $(\overline{15})$ and $(\overline{16a})$ refers to one linearly ligating ligand situated upon the Z axis, we call it $\overline{A}^Z$, then the relation

$$\boldsymbol{F_{ut} = \delta_{ut}} \tag{18}$$

is valid and one obtains for d electrons

$$
\begin{aligned}
\overline{hd\sigma} &= <d\sigma|\overline{A}^Z|d\sigma> = (4/5\, e_\sigma - (2/5)\, e_\pi - (2/5)\, e_\delta \\
\overline{hd\pi} &= <d\pi s|\overline{A}^Z|d\pi s> = <d\pi c|\overline{A}^Z|d\pi c> \\
&= (-1/5\, e_\sigma + (3/5)\, e_\pi - (2/5)\, e_\delta \\
\overline{hd\delta} &= <d\delta s|\overline{A}^Z|d\delta s> = <d\delta c|\overline{A}^Z|d\delta c> \\
&= -(1/5)\, e_\sigma - (2/5)\, e_\pi + (3/5)\, e_\delta
\end{aligned} \tag{19}
$$

corresponding to Eq. $(\overline{15})$, or

$$
\begin{aligned}
\overline{hd\sigma} &= <d\sigma|\overline{A}^Z|d\sigma> = (4/5)\, e'_\sigma - (2/5)\, e'_\pi \\
\overline{hd\pi} &= <d\pi s|\overline{A}^Z|d\pi s> = <d\pi c|\overline{A}^Z|d\pi c> = -(1/5)\, e'_\sigma + (3/5)\, e'_\pi \\
\overline{hd\delta} &= <d\delta s|\overline{A}^Z|d\delta s> = <d\delta c|\overline{A}^Z|d\delta c> = -(1/5)\, e'_\sigma - (2/5)\, e'_\pi
\end{aligned} \tag{20}
$$

corresponding to Eq. $(\overline{16a})$. In both Eqs. (19) and (20) it is seen that the sum of the energies within the d shell is vanishing, as it should be for the barred operator. The h symbols to the extreme left of Eqs. (19) and (20) signify that the diagonal elements are eigenvalues, i.e. that all non-diagonal elements are vanishing.

One may also go back to Eq. (7) and obtain the results of Eqs. $(\overline{15})$ and $(\overline{16a})$ directly for differences between diagonal elements and for non-diagonal elements

$$
\begin{aligned}
<u|\overline{A}^k|u> - <v|\overline{A}^k|v> &= <u|A^k|u> - <v|A^k|v> \\
&= \sum_{t=\sigma,\pi,\delta} [(F_{ut}^k)^2 - (F_{vt}^k)^2]\, e_t \\
&= \sum_{t=\sigma,\pi} [(F_{ut}^k)^2 - (F_{vt}^k)^2]\, e'_t .
\end{aligned} \tag{21}
$$

Table 3. The symmetrical matrix $\mathbf{A}_{d'|e_\sigma}^k$ (only elements on and above the diagonal are given). The elements of this symmetric matrix are the coefficients to $(e_\sigma - e_\delta) = e'_\sigma = e'_\delta$ of Eq. (16a). They correspond to position k of the (not necessarily linearly ligating) ligand given by the direction cosines (α, β, γ) referred to the basic space-fixed coordinate system XYZ, relative to which the real (unprimed) d orbitals are defined. δ has been written as an abbreviation for $+\sqrt{\alpha^2 + \beta^2}$

$\mathbf{A}_{d'|e_\sigma}^k$ $= \mathbf{A}_{d'|e_\sigma}^k - [1/(2l+1)]\,\mathbf{1}$		$\begin{array}{c}\|(z^2)\rangle \\ \|(\sigma)\rangle \\ \|0\rangle\end{array}$	$\begin{array}{c}\|(yz)\rangle \\ \|(\pi s)\rangle \\ \|1\rangle\end{array}$	$\begin{array}{c}\|(zx)\rangle \\ \|(\pi c)\rangle \\ \|2\rangle\end{array}$	$\begin{array}{c}\|(xy)\rangle \\ \|(\delta s)\rangle \\ \|3\rangle\end{array}$	$\begin{array}{c}\|(x^2-y^2)\rangle \\ \|(\delta c)\rangle \\ \|4\rangle\end{array}$
$\langle(z^2)\|$	$\langle 0\|$	$(1/4)(2\gamma^2-\delta^2)^2 -(1/5)$	$(\sqrt{3}/2)(\beta\gamma)(2\gamma^2-\delta^2)$	$(\sqrt{3}/2)(\gamma\alpha)(2\gamma^2-\delta^2)$	$(\sqrt{3}/2)(\alpha\beta)(2\gamma^2-\delta^2)$	$(\sqrt{3}/4)(\alpha^2-\beta^2)(2\gamma^2-\delta^2)$
$\langle(yz)\|$	$\langle 1\|$		$3\beta^2\gamma^2 -(1/5)$	$3\alpha\beta\gamma^2$	$3\alpha\beta^2\gamma$	$(3/2)(\alpha^2-\beta^2)\beta\gamma$
$\langle(zx)\|$	$\langle 2\|$			$3\gamma^2\alpha^2 -(1/5)$	$3\alpha^2\beta\gamma$	$(3/2)(\alpha^2-\beta^2)\gamma\alpha$
$\langle(xy)\|$	$\langle 3\|$				$3\alpha^2\beta^2 -(1/5)$	$(3/2)(\alpha^2-\beta^2)\alpha\beta$
$\langle(x^2-y^2)\|$	$\langle 4\|$					$(3/4)(\alpha^2-\beta^2)^2 -(1/5)$

Table 4. *(see text for Tables 3 and 5)*

$$\bar{A}^{k}_{\pi s}/e_{\pi s} = A^{k}_{\pi s}/e_{\pi s} - [1/(2l+1)]\,\mathbf{1}$$

		$\|(z^2)>$ $\|(\sigma)>$ $\|0>$	$\|(yz)>$ $\|(\pi s)>$ $\|1>$	$\|(zx)>$ $\|(\pi c)>$ $\|2>$	$\|(xy)>$ $\|(\delta s)>$ $\|3>$	$\|(x^2-y^2)>$ $\|(\delta c)>$ $\|4>$
$<(z^2)\|$	$<(\sigma)\|$ $<0\|$	$-(1/5)$	0	0	0	0
$<(yz)\|$	$<(\pi s)\|$ $<1\|$		$\gamma^2\alpha^2/\delta^2 - (1/5)$	$-\alpha\beta\gamma^2/\delta^2$	$\gamma\alpha(\alpha^2-\beta^2)/\delta^2$	$-2\alpha^2\beta\gamma/\delta^2$
$<(zx)\|$	$<(\pi c)\|$ $<2\|$			$\beta^2\gamma^2/\delta^2 - (1/5)$	$-\beta\gamma(\alpha^2-\beta^2)/\delta^2$	$2\alpha\beta^2\gamma/\delta^2$
$<(xy)\|$	$<(\delta s)\|$ $<3\|$				$(\alpha^2-\beta^2)^2/\delta^2 - (1/5)$	$-2\alpha\beta(\alpha^2-\beta^2)/\delta^2$
$<(x^2-y^2)\|$	$<(\delta c)\|$ $<4\|$					$4\alpha^2\beta^2/\delta^2 - (1/5)$

Table 5. The symmetrical matrix $\mathbf{A}^k_{\pi c}|e_{\pi c}$ (only elements on and above the diagonal are given). This matrix is analogous to those of Tables 3 and 4 and is valid also for non-linearly ligating ligands. When the ligands are linearly ligating, $e_{\pi s}=e_{\pi c}=e_\pi$ and the coefficients to $e_\pi-e_\delta=e'_\pi$ of Eq. $(\overline{16a})$ are found by adding the matrices of Tables 4 and 5

$\mathbf{A}^k_{\pi c}|e_{\pi c}$ $=\mathbf{A}^k_{\pi c}|e_{\pi c}-[1/(2l+1)]\,\mathbf{1}$		$\|z^2\rangle$ $\|(\sigma)\rangle$ $\|0\rangle$	$\|(yz)\rangle$ $\|(\pi s)\rangle$ $\|1\rangle$	$\|(zx)\rangle$ $\|(\pi c)\rangle$ $\|2\rangle$	$\|(xy)\rangle$ $\|(\delta s)\rangle$ $\|3\rangle$	$\|(x^2-y^2)\rangle$ $\|(\delta c)\rangle$ $\|4\rangle$
$\langle(z^2)\|$	$\langle 0\|$	$3\gamma^2\delta^2-(1/5)$	$-\sqrt3\beta\gamma(\gamma^2-\delta^2)$	$-\sqrt3\gamma\alpha(\gamma^2-\delta^2)$	$-2\sqrt3\alpha\beta\gamma^2$	$-\sqrt3\gamma^2(\alpha^2-\beta^2)$
$\langle(yz)\|$	$\langle 1\|$		$\beta^2(\gamma^2-\delta^2)^2/\delta^2$ $-(1/5)$	$\alpha\beta(\gamma^2-\delta^2)^2/\delta^2$	$2\alpha\beta^2\gamma(\gamma^2-\delta^2)/\delta^2$	$\beta\gamma(\alpha^2-\beta^2)(\gamma^2-\delta^2)/\delta^2$
$\langle(zx)\|$	$\langle 2\|$			$\alpha^2(\gamma^2-\delta^2)^2/\delta^2$ $-(1/5)$	$2\alpha^2\beta\gamma(\gamma^2-\delta^2)/\delta^2$	$\gamma\alpha(\alpha^2-\beta^2)(\gamma^2-\delta^2)/\delta^2$
$\langle(xy)\|$	$\langle 3\|$				$4\alpha^2\beta^2\gamma^2/\delta^2-(1/5)$	$2\alpha\beta\gamma^2(\alpha^2-\beta^2)/\delta^2$
$\langle(x^2-y^2)\|$	$\langle 4\|$					$(\alpha^2-\beta^2)^2\gamma^2/\delta^2-(1/5)$

From the expression after the second equality sign of Eq. (21) one sees that the coefficients to e_t summed over t vanish, since each of the squared terms adds up to unity. The expression after the third equality sign may then be obtained analogously to Eq. ($\overline{16a}$), and as a consequence of Eq. (10a) the expression

$$<u|\bar{A}^k|v> = \sum_{t=\sigma,\pi} F_{ut}\, F_{vt}\, e_t' \tag{22}$$

is valid for $u \neq v$. Eqs. (21) and (22) mean that, if only energy differences and non-diagonal elements are considered, the zero-point corrected angular-overlap model (the barred operator and barred coefficients) is no longer necessary. It is enough to delete the unnecessary parameter. This property, particularly as it relates to energy differences, such as occur in Eq. (21), has been used previously (12, 13) to translate results of the angular-overlap model into those of the electrostatic model.

For p electrons, the expressions analogous to Eqs. (19) and (20) are

$$\overline{hp\sigma} = <p\sigma|\bar{A}^Z|p\sigma> = (2/3)e_\sigma(\mathrm{p}) - (2/3)e_\pi(p)$$
$$\overline{hp\pi} = <p\pi s|\bar{A}^Z|p\pi s> = <p\pi c|\bar{A}^Z|p\pi c> \tag{23}$$
$$= -(1/3)e_\sigma(p) + (1/3)e_\pi(p)$$

and

$$\overline{hp\sigma} = <p\sigma|\bar{A}^Z|p\sigma> = (2/3)e_\sigma'(p)$$
$$\overline{hp\pi} = <p\pi s|\bar{A}^Z|p\pi s> = <p\pi c|\bar{A}^Z|p\pi c> = -(1/3)e_\sigma'(p) \tag{24}$$

where we have a p in parenthesis to distinguish the energy parameters here from those for d electrons.

It is seen that, for linearly ligating ligands acting on p electron systems, there is only one semiempirical AOM parameter

$$e_\sigma'(p) = e_\sigma(p) - e_\pi(p) \tag{25}$$

The fact that AOM in this case requires only one parameter does not imply that it does not allow for the p electrons of the central ion to take part in π bonding with the ligand. The reason is that the symmetry, because of the linear ligation, is so high that only one energy difference is present with the p shell. We have a situation analogous to that for d electrons in an octahedron where (9)

$$\Delta(O_h) \equiv hde(O_h) - hdt_2(O_h) = (3e_\sigma + 3e_\delta) - (4e_\pi + 2e_\delta)$$
$$= 3e_\sigma' - 4e_\pi' \tag{26}$$

is the only observable energy difference so that e'_σ and e'_π cannot be determined individually from experiments on systems of full octahedral symmetry.

4. The Angular-Overlap Model for Linearly Ligating Ligands. Axial Spherical Harmonic and Angular Overlap Symmetry Parameterizations

It has been shown (10) that the angular overlap model can be regarded as a first-order perturbation model. For the angular-overlap model representing one linearly ligating ligand upon the Z axis the perturbation operator may be expanded into terms belonging to irreducible representations of the three-dimensional rotation group. These terms must further be totally symmetric under the group $C_{\infty v}$ of the diatomic entity, which is another way of stating that they must have σ symmetry around the Z axis. Thus the potential $\bar{A}^Z$ representing this single ligand, as in Eqs. (19) and (20), may be expressed as

$$\bar{A}^Z = \sum_{L>0} \sum_{T=0}^{T=2L} \mathfrak{A}_T^L \, \mathfrak{H}_T^L = \sum_{L>0} \mathfrak{A}_0^L \, \mathfrak{H}_0^L \tag{27}$$

$$= \sum_{L>0} \mathfrak{A}_0^L \, r^L \, \mathfrak{C}_0^L = \sum_{L>0} \mathfrak{A}_0^L \, r^L \, P_L(\cos\theta)$$

where $\mathfrak{A}_T^L$ is the potential parameter and $\mathfrak{H}_T^L$ the solid spherical harmonic [normalized to $4\pi/(2L+1)$] of the degree L, T characterizing the particular component of the irreducible tensorial set (6, 15).

$\mathfrak{A}_T^L$ vanishes for $T \neq 0$, because the Z axis is an axis of symmetry. $\mathfrak{C}_0^L = P_L(\cos\theta)$ is the Legendre polynomial or axial spherical harmonic and r is the absolute value of radius vector of the electron. The term $L=0$ does not appear in Eq. (27) because for this term the barycenter rule does not apply, as it ought to, since we are concerned with the barred operator [cf. Ref. (15), legend to Table 16].

Using the expansion of Eq. (27) for $\bar{A}^Z$, the general matrix element is

$$<\alpha lu|\bar{A}^Z|\alpha lv> = \sum_{L>0} <R(\alpha l)|\mathfrak{A}_0^L \, r^L|R(\alpha l)> \, <lu|\mathfrak{C}_0^L|lv>$$

$$= \sum_{L>0} I^L <lu|\mathfrak{C}_0^L|lv> \tag{28}$$

$$= \sum_{L>0} I^L <l||\mathfrak{C}^L||l> \begin{pmatrix} l & L & l \\ u & 0 & v \end{pmatrix}$$

$$= \sum_{L>0} E^L \begin{pmatrix} l & L & l \\ u & 0 & v \end{pmatrix}$$

where

$$\begin{pmatrix} l & L & l \\ u & 0 & v \end{pmatrix} = \begin{pmatrix} l & L & l \\ u & \sigma & v \end{pmatrix} \tag{29}$$

is a 3-l symbol $(6, 15)$ and where the reduced matrix element of $\mathfrak{C}^L$, $<l||\mathfrak{C}^L||l>$, is given [Ref. (6), Eqs. (21) and (41)] by

$$<l||\mathfrak{C}^L||l> = (2l+1) \begin{pmatrix} l & L & l \\ 0 & 0 & 0 \end{pmatrix} = (2l+1) \begin{pmatrix} l & L & l \\ \sigma & \sigma & \sigma \end{pmatrix}$$

$$= (2l+1) \frac{[l+(L/2)]! \, L!}{[l-(L/2)]! \, [(L/2)\,!]^2} \sqrt{\frac{(2l-L)!}{(2l+L+1)!}} \tag{30}$$

where in Eq. (30) the special conditions, $l_1 = l_3 = l$ and $l_2 = L$, have been introduced into the general expression [Ref. (6), Eq. (21)]

$$\begin{pmatrix} l_1 & l_2 & l_3 \\ \sigma & \sigma & \sigma \end{pmatrix} = [(l_1+l_2+l_3)/2]! \, \{[(l_2+l_3-l_1)/2]! \, [(l_3+l_1-l_2)/2]!$$

$$\times [(l_1+l_2-l_3)/2]! \,\}^{-1}\{(l_2+l_3-l_1)! \, (l_3+l_1-l_2)! \tag{31}$$

$$\times (l_1+l_2-l_3)!\}^{\frac{1}{2}} \, (l_1+l_2+l_3+1)^{-\frac{1}{2}}$$

The axial spherical harmonic parameters I^L and E^L are the empirical parameters, which contain the unknown potential parameter $\mathfrak{A}^L$, integrals over the expanded radial function, characterized by α, and in the case of the so-called (6) reduced ligand-field parameter E^L, also the symmetry-determined quantity $<l||\mathfrak{C}^L||l>$ given in Eq. (30).

It is particularly important to compare the third and the fifth expressions of Eq. (28). This comparison, which may be expressed in words by saying that $<lu|\mathfrak{C}_0^L|lv> = <lu|\mathsf{P}_L|lv>$ is proportional to the 3-l symbol $\begin{pmatrix} l & L & l \\ u & 0 & v \end{pmatrix}$, has recently (15), by choosing this constant of proportionality positive, led to a phase standardization of the even 3-l symbols. In Appendix 2 an integration by differentiation method [Ref. (18) p. 156] has been developed for evaluating integrals of the type $<lu|\mathfrak{C}_T^L|lv>$, as well as even 3-l symbols of the type $\begin{pmatrix} l_1 & l_2 & l_3 \\ t_1 & t_2 & t_3 \end{pmatrix}$.

The coefficients to I^L can now either be found by evaluating the integrals $<lu|\mathfrak{C}_0^L|lv>$ or the 3-l symbols $\begin{pmatrix} l & L & l \\ u & 0 & v \end{pmatrix}$ directly (Appendix 2), or by using Eqs. (28) and (30) together with a table of 3-l symbols (6). The results for $l=1$ and $l=2$ are

$$\overline{hp\sigma} = <p\sigma|\bar{\mathsf{A}}^z|p\sigma> = \qquad\qquad + 2/5 \, I^2(p)$$

$$\overline{hp\pi} = <p\pi s|\bar{\mathsf{A}}^z|p\pi s> = <p\pi c|\bar{\mathsf{A}}^z|p\pi c> = -1/5 \, I^2(p) \tag{32}$$

$$\overline{hd\sigma} = <d\sigma|\bar{A}^z|d\sigma> = \qquad\qquad\qquad + 2/7\ I^2(d) + 6/21\ I^4(d)$$
$$\overline{hd\pi} = <d\pi s|\bar{A}^z|d\pi s> = <d\pi c|\bar{A}^z|d\pi c> = + 1/7\ I^2(d) - 4/21\ I^4(d)$$
$$\overline{hd\delta} = <d\delta s|\bar{A}^z|d\delta s> = <d\delta c|\bar{A}^z|d\delta c> = - 2/7\ I^2(d) + 1/21\ I^4(d)$$

$$(33)$$

The integration can also be done using surface harmonics.

For $|p\pi s>$ we have, for example, for the coefficient to I^2

$$\int_0^{2\pi} p\pi s(\varphi)\ p\pi s(\varphi)\ d\varphi \times \int_0^{\pi} p\pi(\theta)\ \mathbf{P}_2\ (\cos\theta)\ p\pi(\theta)\ \sin\theta\ d\theta$$
$$= \int_0^{\pi} [p\pi(\theta)]^2\ \mathbf{P}_2\ (\cos\theta)\ \sin\theta\ d\theta \qquad\qquad (34)$$

Inserting now $\mathbf{P}_2 = (1/2)\ (3\cos^2\theta - 1)$ and $p\pi(\theta) = (\sqrt{3}/2)\ \sin\theta$, the last integral is calculated as $-(1/5)$, as stated in Eq. (32). A few comments may be added on, for example, Eq. (32). The L values are limited by the fact that the direct product of two p spaces in product space only gives the three gerade terms, s, p, and d, of which the p term vanishes because of the ungerade character of $\mathbf{P}_1(\cos\theta)$ [see also Ref. (6) p. 279]. The degeneracy of the π functions is a consequence of the symmetry $C_{\infty v}$ of the system, consisting of the central ion and the linearly ligating ligand. Further it may be noted that in Eq. (34) $|p\pi s>$ has been written as

$$|p\pi s> = p\pi s(\varphi)\ p\pi(\theta) \qquad\qquad (35)$$

where each factor separately is normalized to unity.

The weighted one-electron energies of Eqs. (32) and (33) add up to zero, and a comparison of Eqs. (32) and (24) for p electrons and of Eqs. (33) and (20) for d electrons gives one-to-one relationships between the two parameterizations[6].

---

[6] With reference to the comments on Eq. (27), it may be worth noting that the relationship between the two equivalent parameterizations of the angular overlap model for linearly ligating ligands may also be evaluated on the basis of the unbarred operator $\mathbf{A}^z$. For d electrons we then obtain the matrix equation

$$\begin{bmatrix} e_\sigma + e(r) \\ e_\pi + e(r) \\ e_\delta + e(r) \end{bmatrix} = \begin{bmatrix} 1 & 2/7 & 6/21 \\ 1 & 1/7 & -4/21 \\ 1 & -2/7 & 1/21 \end{bmatrix} \begin{bmatrix} I^0 \\ I^2 \\ I^4 \end{bmatrix} \qquad (37a)$$

or the reciprocal one

$$\begin{bmatrix} I^0 \\ I^2 \\ I^4 \end{bmatrix} = \begin{bmatrix} 1/5 & 2/5 & 2/5 \\ 1 & 1 & -2 \\ 9/5 & -12/5 & 3/5 \end{bmatrix} \begin{bmatrix} e_\sigma + e(r) \\ e_\pi + e(r) \\ e_\delta + e(r) \end{bmatrix} \qquad (37b)$$

It is characteristic of the coefficient matrix of Eq. (37b) that the sums of the elements within the second and third rows are equal to zero. This has the consequence that $e(r)$ vanishes in expressions for I^2 and I^4 and at the same time the translation into the primed parameters, e'_σ and e'_π, becomes possible.

In view of the interpretation of the e parameters of the angular-overlap model, a significant way of obtaining this relationship is to express the energies of the $l\lambda$ orbitals ($\lambda < l$), using the energy of the $l\lambda$ orbitals ($l = \lambda$) as zero point. One then obtains for p electrons

$$<p\sigma|\bar{A}^Z|p\sigma> - <p\pi|\bar{A}^Z|p\pi> = e'_\sigma(p) = 3/5\, I^2(p) \qquad (36a)$$

and for d electrons

$$
\begin{aligned}
<d\sigma|\bar{A}^Z|d\sigma> - <d\delta|\bar{A}^Z|d\delta> &= e'_\sigma = (4/7)\, I^2 + (5/21)\, I^4 \\
<d\pi|\bar{A}^Z|d\pi> - <d\delta|\bar{A}^Z|d\delta> &= e'_\pi = (3/7)\, I^2 - (5/21)\, I^4
\end{aligned}
\qquad (\overline{37a})
$$

or the reverse relations, for p electrons,

$$I^2(p) = (5/3)\, e'_\sigma(p) \qquad (36b)$$

and for d electrons

$$
\begin{aligned}
I^2 &= e'_\sigma + e'_\pi \\
I^4 &= (3/5)\, (3e'_\sigma - 4e'_\pi)
\end{aligned}
\qquad (\overline{37b})
$$

Relationships analogous to those of Eqs. (37) have been used previously (12, 13) for the comparison of the angular-overlap model and the point-charge or point-dipole electrostatic model.

When the perturbation from more ligands is to be taken into account, the addition theorem for spherical harmonics is usually used to develop the expression of the electrostatic model when more ligands are involved. This can be done very simply by means of the concepts of the angular-overlap model.

With assumption III (p. 71) the general matrix element using Eqs. (27) and (28) can be written.

$$<du|\bar{A}|dv> = \sum_{k=1}^{k=N} \sum_{L>0}^{L=4} I^L <du|\, \mathsf{P}_L^k\,|dv> \qquad (38)$$

where the integration over the radial functions has been performed. The Legendre polynomial P_L^k, also called the axial harmonic, or the $L\Sigma'$ function, normalized to $4\pi/(2L+1)$, has its axis of symmetry coinciding with the Z' axis corresponding to the ligand k. du and dv are the angular parts of the d functions, normalized to unity. For reasons analogous to those discussed in connection with Eq. (32) these integrals over the product of two d functions involve only P_L terms when L is equal to S, D, and G; the S term is absent here because we are concerned with barred energies. Now the $L\Sigma'$ function can, using Eq. (5), be written

$$(L\Sigma') = \sum_{T=0}^{T=2L} (LT)\, \mathbf{F}_{T\Sigma'}^{(L)} \qquad (39a)$$

Eq. (39 a) is in effect the addition theorem for spherical harmonics. This can be seen by rewriting it as

$$
\begin{aligned}
\mathsf{P}_L^k &= \sum_{T=0}^{T=2L} \mathbb{C}_T^L \, F_{T\Sigma'}^{(L)} \\
&= \sum_{T=0}^{T=2L} \mathbb{C}_T^L \, \mathfrak{H}_T^L(\alpha_k, \beta_k, \gamma_k) = \sum_{T=0}^{T=2L} \mathbb{C}_T^L \, \mathbb{C}_T^L(\theta_k, \varphi_k)
\end{aligned} \tag{39b}
$$

remembering that the σ angular-overlap integral between a central ion orbital $|LT\rangle$ and a ligand orbital, whose ligator has the coordinates $(\alpha_k, \beta_k, \gamma_k)$, is obtained by inserting these coordinates into $|LT\rangle$, renormalized to $4\pi/(2L+1)$; i.e. into $\mathfrak{H}_T^L$. Using the expression of Eq. (39 b) and introducing the group angular-overlap integral

$$
\begin{aligned}
F_{T\,a_1\,(\Sigma)}^{(L)} &= (1/\sqrt{N}) \sum_{k=1}^{k=N} F_{T\Sigma'}^{(L)} \\
&= (1/\sqrt{N}) \sum_{k=1}^{k=N} \mathfrak{H}_T^L(\alpha_k, \beta_k, \gamma_k)
\end{aligned} \tag{40}
$$

between the central ion orbital $|LT\rangle$ and the (normalized) symmetrical linear combination of ligand Σ orbitals, we obtain the expression for the general matrix element of Eq. (38) between the u and v orbitals.

$$
\begin{aligned}
\langle du|\bar{\mathsf{A}}|dv\rangle &= \sum_{L=2,4} I^L \sum_{T=0}^{T=2L} \langle du|\mathbb{C}_T^L|dv\rangle \sum_{k=1}^{k=N} \mathfrak{H}_T^L(\alpha_k, \beta_k, \gamma_k) \\
&= \sum_{L=2,4} I^L \sum_{T=0}^{T=2L} \langle du|\mathbb{C}_T^L|dv\rangle \sqrt{N} \, F_{T\,a_1\,(\Sigma)}^{(L)}
\end{aligned} \tag{41}
$$

The third expression of Eq. (41) is particularly useful for symmetry considerations because, since $a_1(\Sigma)$ belongs to the totally symmetric irreducible representation of the molecular point group, the function $|LT\rangle$, and thereby the operator $\mathbb{C}_T^L$, must contain something totally symmetrical in order to give a non-vanishing group angular-overlap integral. This, for example, allows one to conclude immediately that for any cubic system the $\mathbb{C}_T^L$ terms with $L=2$ must vanish because D functions in no such case span the totally symmetric irreducible representation. This is, of course, a well-known conclusion which one may arrive at directly by requiring the perturbation operator to be totally symmetrical.

Table 6. *The symmetrical matrices of the unit tensorial operators $\mathfrak{B}^D$ and $\mathfrak{B}^G$ with respect to the d basis (only elements on and above the diagonal are given). The unit tensorial operator $\mathfrak{B}$, whose matrix elements are the 3-l symbols, is defined by its reduced matrix [Ref. (15) p. 243] so that in this case*

$$<l\|\mathfrak{B}^L\|l> = \delta(l\,L\,l)$$

i.e. vanishing if the direct triple product does not contain an S term. The 3-l symbols are here the so-called even 3-l symbols [Ref. (15) Sect. 4a], for which $\begin{pmatrix} l & L & l \\ u & T & v \end{pmatrix}$ is proportional to the integral $<lu|\mathfrak{C}_T^L|lv>$ with a positive constant of proportionality. The entities in the $\mathfrak{B}^D$ table should be multiplied by $\sqrt{7|70}$ in order to obtain the 3-l symbols and by $1|7$ to obtain the corresponding integrals. For the $\mathfrak{B}^G$ table the multipliers are $\sqrt{7|(72\times35)}$ and $1|42$, respectively. The components T of $\mathfrak{B}_L$ or $\mathfrak{C}_L$ are given as numbers between vertical lines above the value of the corresponding matrix element. For example,

$$\begin{pmatrix} d & D & d \\ \pi s & \pi s & \delta c \end{pmatrix} = \begin{pmatrix} d & D & d \\ 1 & 1 & 4 \end{pmatrix} = -\sqrt{3}\times\sqrt{1|70} = -\sqrt{3|70}$$

where $\sqrt{1|70}$ is the constant of proportionality mentioned above. – The numbering of the functions is different from that previously used [Ref. (8) p. 371]. Here the symbol for σ functions has been chosen as $|0>$, and that for a λc function as $|2\lambda>$ (a ket with an even number in it), whereas for a λs function $|2\lambda-1>$ (a ket with an odd number in it) has been chosen. This new numbering has several advantages which are apparent from the tables. The following sum rules should be noted [Ref. (15) p. 251]

$$\sum_{u,T,v} \begin{pmatrix} l & L & l \\ u & T & v \end{pmatrix}^2 = 1$$

$$\sum_{u,v} \begin{pmatrix} l & L & l \\ u & T & v \end{pmatrix}^2 = 1/(2L+1)$$

$$\sum_{u,T} \begin{pmatrix} l & L & l \\ u & T & v \end{pmatrix}^2 = \sum_{v,T} \begin{pmatrix} l & L & l \\ u & T & v \end{pmatrix}^2 = 1/(2l+1)$$

$$\sum_{u} \begin{pmatrix} l & L & l \\ u & T & u \end{pmatrix} = 0; \; L \neq S$$

$\mathfrak{B}^D$

	$\begin{array}{c}\|(z^2)\rangle\\\|(\sigma)\rangle\\\|0\rangle\end{array}$	$\begin{array}{c}\|(yz)\rangle\\\|(\pi s)\rangle\\\|1\rangle\end{array}$	$\begin{array}{c}\|(zx)\rangle\\\|(\pi c)\rangle\\\|2\rangle\end{array}$	$\begin{array}{c}\|(xy)\rangle\\\|(\delta s)\rangle\\\|3\rangle\end{array}$	$\begin{array}{c}\|(x^2-y^2)\rangle\\\|(\delta c)\rangle\\\|4\rangle\end{array}$
$\langle(\sigma)\|\ \langle 0\| \quad \langle(z^2)\|$	$\dfrac{\|0\|}{2}$	$\dfrac{\|1\|}{1}$	$\dfrac{\|2\|}{1}$	$\dfrac{\|3\|}{-2}$	$\dfrac{\|4\|}{-2}$
$\langle(\pi s)\|\ \langle 1\| \quad \langle(yz)\|$			$\dfrac{\|3\|}{\sqrt3}$	$\dfrac{\|2\|}{\sqrt3}$	$\dfrac{\|1\|}{-\sqrt3}$
$\langle(\pi c)\|\ \langle 2\| \quad \langle(zx)\|$			$\dfrac{\|4\|}{\sqrt3}$	$\dfrac{\|1\|}{\sqrt3}$	$\dfrac{\|2\|}{\sqrt3}$
$\langle(\delta s)\|\ \langle 3\| \quad \langle(xy)\|$		$\dfrac{\|4\|}{-\sqrt3}$	$\dfrac{\|0\|}{1}$		0
$\langle(\delta c)\|\ \langle 4\| \quad \langle(x^2-y^2)\|$		$\dfrac{\|0\|}{1}$		$\dfrac{\|0\|}{-2}$	$\dfrac{\|0\|}{-2}$

$\mathfrak{B}^G$

	$\begin{array}{c}\|(z^2)\rangle\\\|(\sigma)\rangle\\\|0\rangle\end{array}$	$\begin{array}{c}\|(yz)\rangle\\\|(\pi s)\rangle\\\|1\rangle\end{array}$	$\begin{array}{c}\|(zx)\rangle\\\|(\pi c)\rangle\\\|2\rangle\end{array}$	$\begin{array}{c}\|(xy)\rangle\\\|(\delta s)\rangle\\\|3\rangle\end{array}$	$\begin{array}{c}\|(x^2-y^2)\rangle\\\|(\delta c)\rangle\\\|4\rangle\end{array}$
$\langle(\sigma)\|\ \langle 0\| \quad \langle(z^2)\|$	$\dfrac{\|0\|}{12}$	$\dfrac{\|1\|}{2\sqrt{30}}$	$\dfrac{\|2\|}{2\sqrt{30}}$	$\dfrac{\|3\|}{2\sqrt{15}}\ ;\ \dfrac{\|5\|}{\sqrt{70}}$	$\dfrac{\|4\|}{2\sqrt{15}}\ ;\ \dfrac{\|6\|}{\sqrt{70}}$
$\langle(\pi s)\|\ \langle 1\| \quad \langle(yz)\|$			$\dfrac{\|3\|}{2\sqrt{20}}$	$\dfrac{\|2\|}{-\sqrt{10}}\ ;\ \dfrac{\|6\|}{-\sqrt{70}}$	$\dfrac{\|1\|}{\sqrt{10}}\ ;\ \dfrac{\|5\|}{\sqrt{70}}$
$\langle(\pi c)\|\ \langle 2\| \quad \langle(zx)\|$		$\dfrac{\|4\|}{-2\sqrt{20}}$	$\dfrac{\|4\|}{2\sqrt{20}}$	$\dfrac{\|1\|}{-\sqrt{10}}\ ;\ \dfrac{\|5\|}{\sqrt{70}}$	$\dfrac{\|2\|}{-\sqrt{10}}\ ;\ \dfrac{\|8\|}{2\sqrt{35}}$
$\langle(\delta s)\|\ \langle 3\| \quad \langle(xy)\|$		$\dfrac{\|0\|}{-8}$	$\dfrac{\|0\|}{-8}$	$\dfrac{\|0\|}{2}\ ;\ \dfrac{\|8\|}{-2\sqrt{35}}$	$\dfrac{\|7\|}{2\sqrt{35}}$
$\langle(\delta c)\|\ \langle 4\| \quad \langle(x^2-y^2)\|$					$\dfrac{\|0\|}{2}\ ;\ \dfrac{\|8\|}{2\sqrt{35}}$

Table 7. *The two sets of solid spherical harmonics $\mathfrak{H}^l$ ($l = 2$ and $l = 4$), appearing in the integrals $<d\,|\mathfrak{H}^l|\,d>$, $\mathfrak{H}^l$ are normalized to $4\pi/(2l + 1)$ over the unit sphere*

$$\mathfrak{H}^d_\sigma = (1/2)\,(2z^2 - x^2 - y^2)$$
$$\mathfrak{H}^d_{\pi s} = \sqrt{3}\,zy$$
$$\mathfrak{H}^d_{\pi c} = \sqrt{3}\,zx$$
$$\mathfrak{H}^d_{\delta s} = \sqrt{3}\,xy$$
$$\mathfrak{H}^d_{\delta c} = (\sqrt{3}/2)\,(x^2 - y^2)$$
$$\mathfrak{H}^g_\sigma = (1/8)\,(8z^4 + 3x^4 + 3y^4 + 6x^2y^2 - 24z^2y^2 - 24z^2x^2)$$
$$\mathfrak{H}^g_{\pi s} = (\sqrt{10}/4)\,(4z^3y - 3zx^2y - 3zy^3)$$
$$\mathfrak{H}^g_{\pi c} = (\sqrt{10}/4)\,(4z^3x - 3zx^3 - 3zxy^2)$$
$$\mathfrak{H}^g_{\delta s} = (\sqrt{5}/2)\,(6z^2xy - x^3y - xy^3)$$
$$\mathfrak{H}^g_{\delta c} = (\sqrt{5}/4)\,(6z^2x^2 - 6z^2y^2 - x^4 + y^4)$$
$$\mathfrak{H}^g_{\varphi s} = (\sqrt{70}/4)\,(3zx^2y - zy^3)$$
$$\mathfrak{H}^g_{\varphi c} = (\sqrt{70}/4)\,(zx^3 - 3zxy^2)$$
$$\mathfrak{H}^g_{\gamma s} = (\sqrt{35}/2)\,(x^3y - xy^3)$$
$$\mathfrak{H}^g_{\gamma c} = (\sqrt{35}/8)\,(x^4 - 6x^2y^2 + y^4)$$

Eq. (41) factorizes the coefficient to the semiempirical parameter I^L as a first factor, $<du\,|\mathfrak{C}^L_T|\,dv>$, tabulated in Table 6, and depending only on the symmetry of the system. This factor, the symmetry coefficient, may be calculated using Table 7 and Eq. (98). The second factor, the geometry coefficient $\sqrt{N}\,F^{(L)}_{T\,\mathbf{a}_1\,(\Sigma)}$, depends on the coordinates of the ligands (α_k, β_k, γ_k) and hence on the detailed geometry of the system[7]. Often the ligand-field model is used without assumption III of p. 71. In this case one may call it a non-additivity model and the geometry coefficient is then included in I^L, which then may take different values for different T and therefore by analogy with Eq. (28) may be called I^L_T.

We then have

$$
\begin{aligned}
<du\,|\nabla|\,dv> &= \sum_{L,T} I^L_T\,<du\,|\mathfrak{C}^L_T|\,dv> \\
&= \sum_{L,T} E^L_T \begin{pmatrix} d & L & d \\ u & T & v \end{pmatrix}
\end{aligned}
\tag{42}
$$

where the ligand-field operator now has been called ∇.

---

[7] For convenience we use the terms symmetry coefficient and geometry coefficient. However, it should be noted that the geometry coefficient is sometimes also symmetry-determined. This is the case in our cubic example of p. 94 and in all of Table 8.

In Eq. (42) the parameters E_T^L occur. These parameters have been proposed as a rational standard for such symmetry parameters [Ref. (6) p. 279] and have been named reduced ligand-field parameters, even though they depend on T.

Having derived the parameter E_T^L by experiment, and knowing the detailed geometry, it is possible by equating $\bar{A}$ and ∇ to set up the linear equations which connect E_T^L with the parameters e'_σ, e'_π or I^2, I^4 of the additivity model (6). Here it should be noted that the parameters of the additivity model depend on the kind of ligand and, when different ligands are present, must be given an extra index.

So, whereas the relationship between the parameters e'_σ, e'_π on the one side and the parameters I^2, I^4 on the other is a one-to-one relationship, this is in general not true for the relationship between a non-additivity model and an additivity model. Here one or the other model may have the larger number of parameters, depending on the symmetry and on the number of different ligands (or equal ligands with different distances), each of which provides two parameters for the additivity model.

As an illustration of Eq. (41) we consider an octahedral (O_h) and a tetrahedral (T_d) system.

The symmetry factor $<du|\mathfrak{C}_T^L|dv> = <du|\mathfrak{H}_T^L|dv>$ [cf. comment on Eq. (98)] is determined by using for $\mathfrak{H}_T^L$ the harmonic of Eq. (81) which is totally symmetric in both symmetries. Using Eq. (98), one obtains[8] for $u = v = \delta s$

$$<d\delta s|\mathfrak{H}_{a_1}^4|d\delta s> = \frac{5\,4!\,2^4}{0!\,9!}\,\frac{\sqrt{21}}{6}$$
$$\times \left(-3\,\frac{\partial^4}{\partial x^2\,\partial y^2}\right) \nabla^0\,(3x^2\,y^2) = -\frac{2\sqrt{21}}{63} \tag{44}$$

The geometry factor $\sum_{k=1}^{k=N} \mathfrak{H}_{a_1}^4(\alpha_k, \beta_k, \gamma_k)$ is obtained by inserting the ligand coordinates on the unit sphere into $\mathfrak{H}_{a_1}^4$ of Eq. (81). Thereby, apart from the normalization constant $(\sqrt{21}/6)$, each ligand contributes unity to the geometry factor of the octahedron, so that the factor in this case is 6. Choosing for the tetrahedron one ligand as $(\sqrt{1/3}, \sqrt{1/3}, \sqrt{1/3})$, each ligand gives here $(-6/9)$, so that the factor in this case is $(-24/9)$. The ratio between the factor for the tetrahedron and that for

---

[8] Using Table 7 and, for example, the projection operator set $O^{(4)}$ of Eq. (82), $\mathfrak{H}^4(a_1)$ of Eq. (81) may be written

$$\mathfrak{H}_{a_1}^4 = (\sqrt{21}/6)\,\mathfrak{H}_\sigma^4 + (\sqrt{15}/6)\,\mathfrak{H}_{\gamma c}^4 \tag{43}$$

so that Table 6 can be used to derive the integral of Eq. (44).

the octahedron is the well-known (19) one of $-(4/9)$. It may be noted that the ratio here has been obtained by looking at the corresponding matrix elements $<d\delta s|\bar{A}(O_h)|d\delta s>$ and $<d\delta s|\bar{A}(T_d)|d\delta s>$, introducing the additivity model (16), i.e. assumption III of p. 71. It is important that the barred operators be used here. Otherwise only differences between matrix elements [see Eq. (21)] could have been used for such a comparison.

For later reference we note that

$$<d\delta s|\bar{A}(O_h)|d\delta s> = -\frac{2\sqrt{21}}{63} \times \frac{\sqrt{21}}{6} \times 6 \times I^4 = -\frac{2}{3} I^4 \qquad (45)$$

where the first factor is the symmetry factor of Eq. (44), and the second and third factors make up together the geometry coefficient mentioned in connection with Eq. (41).

5. Examples of the Interrelation between the two Parameterizations. Discussion of the Crystal-Field Model

In a previous paper (9) the angular-overlap model, based upon the operator A of Eq. (8), apart from the term $e(r)$, was applied to a series of symmetrical chromophores. In Table 8 the results are given for the same chromophores, based upon the barred operator $\bar{A}$ of Eq. (16b).

Table 8 can be obtained from Table 4 of Ref. (9), using Eqs. (8) and (8) or (16b). In this way one obtains, for example, for the case D_{3h} (3 ligands), ignoring the term $3 e(r)$ of Eq. (8) which was not included in Ref. (9):

$$<d\delta s|A(D_{3h})|d\delta s> = \frac{9}{8} e_\sigma + \frac{3}{2} e_\pi + \frac{3}{8} e_\delta \qquad (46)$$

and from Eqs. (8), (13) and (14) and finally (16b), using the original result stated in Eq. (46):

$$<d\delta s|\bar{A}(D_{3h})|d\delta s> = \left(-\frac{3}{5} + \frac{9}{8}\right)e_\sigma + \left(-\frac{6}{5} + \frac{3}{2}\right)e_\pi + \left(-\frac{6}{5} + \frac{3}{8}\right)e_\delta$$

$$= \frac{21}{40} e'_\sigma + \frac{12}{40} e'_\pi \qquad (47)$$

as stated in Table 8.

Table 8 can alternatively be obtained by calculation using Tables 3, 4 and 5. We repeat the previous example taking as coordinates for the three ligands $(1, 0, 0)$, $\left(-\frac{1}{2}, \frac{\sqrt{3}}{2}, 0\right)$, and $\left(-\frac{1}{2}, -\frac{\sqrt{3}}{2}, 0\right)$. These are the Cartesian coordinates (x, y, z) of the ligands radially projected upon the unit sphere, or its direction cosines (α, β, γ). From Table 3 one obtains

$$<d\delta s|\bar{A}_\sigma(D_{3h})|d\delta s> = \sum_{k=1}^{k=3} [3\,\alpha_k^2\,\beta_k^2 - (N/5)]e_\sigma' = \frac{21}{40}\,e_\sigma' \qquad (48)$$

and from Tables 4 and 5

$$
\begin{aligned}
<d\delta s|\bar{A}_\pi(D_{3h})|d\delta s> &= <d\delta s|\bar{A}_{\pi s}(D_{3h})|d\delta s> + <d\delta s|\bar{A}_{\pi c}(D_{3h})|d\delta s> \\
&= \sum_{k=1}^{k=3} [\alpha_k^2 - \beta_k^2)^2/(\alpha_k^2 + \beta_k^2) \\
&\quad + 4\,\alpha_k^2\,\beta_k^2\,\gamma_k^2/(\alpha_k^2 + \beta_k^2) - (2N/5)] \qquad (49) \\
&= \frac{12}{40}\,e_\pi'
\end{aligned}
$$

in agreement with Eq. (47) and Table 8 through the relation

$$\bar{A} = \bar{A}_\sigma + \bar{A}_\pi \qquad (50)$$

of Eqs. (8) and ($\overline{16b}$).

With Table 8 at hand, one may go over into the reparameterized form of the angular-overlap model for linearly ligating ligands and use the parameters I^2 and I^4 by applying the linear relationships of Eq. ($\overline{37a}$).

One example will serve to illustrate the reparameterization. Let us calculate the matrix element $<d\delta s|\bar{A}(O_h)|d\delta s>$ of Eq. (45). We do this by adding the perturbation contributions from the case $D_{\infty h}$ (two ligands on the Z axis) to the contributions from the case D_{4h} (four ligands on the X and Y axes). The results are obtained from Table 8, using also Eq. ($\overline{37a}$),

$$
\begin{aligned}
<d\delta s|\bar{A}(O_h)|d\delta s> &= \left(-\frac{2}{5}\,e_\sigma' - \frac{4}{5}\,e_\pi'\right) + \left(-\frac{4}{5}\,e_\sigma' + \frac{12}{5}\,e_\pi'\right) \\
&= -\frac{6}{5}\,e_\sigma' + \frac{8}{5}\,e_\pi' \qquad (51) \\
&= -\frac{6}{5}\left(\frac{4}{7}\,I^2 + \frac{5}{21}\,I^4\right) + \frac{8}{5}\left(\frac{3}{7}\,I^2 - \frac{5}{21}\,I^4\right) \\
&= -\frac{2}{3}\,I^4
\end{aligned}
$$

in agreement with Eq. (45).

Table 8. The energies in the angular-overlap model of d orbitals in various ligand fields, measured relative to their average energy. The number of linearly ligating ligands is stated in parenthesis after the point group symbols. For all the symmetries of the table the real standard d orbitals transform as irreducible representations (reps) of the point groups in question. The Z axis has been chosen in all cases as the main axis of symmetry. In order to distribute $d\delta s$ and $d\delta c$ upon $b_2(D_{4h})$ and $b_1(D_{4h})$, respectively, the X, Y axes have been chosen to represent the first of the classes of two-fold axes perpendicular to the four-fold axis. The fact that the contributions to the energies from different ligands are additive, also when expressed in e'_λ parameters, must be emphasized. So, for example, if the results of $D_{\infty h}(2)$ and $D_{4h}(4)$ are added, those of the octahedron are obtained, measured relative to the average of the whole d shell of the octahedral case. — The d orbitals have been given as standard real solid spherical harmonics $\sqrt{5/4\pi}\,\mathfrak{H}^2$, normalized to unity over the unit sphere, and have been further characterized by their standard numbering (cf. footnote to p. 71 and text to Table 6), their λ values, and their additional standard specification t of λ components

d orbital	Standard numbering λ		t	$D_{\infty h}(2)$ linear			$D_{3h}(3)$ triangular			$D_{4h}(4)$ square			$D_{5h}(5)$ pentagonal		
				reps	e'_σ	e'_π	reps	e'_σ	e'_π	reps	e'_σ	e'_π	reps	e'_σ	e'_π
$\sqrt{\frac{5}{4\pi}}\left[z^2-\left(\frac{1}{2}\right)x^2-\left(\frac{1}{2}\right)y^2\right]$	0	0	σ	A_{1g}	$\frac{8}{5}$	$\frac{4}{5}$	A'_1	$\frac{3}{20}$	$-\frac{24}{20}$	A_{1g}	$\frac{1}{5}$	$-\frac{8}{5}$	A'_1	$\frac{1}{4}$	-2
$\sqrt{\frac{5}{4\pi}}\sqrt{3}\,yz$	1	1	πs	E_{1g}	$-\frac{2}{5}$	$\frac{6}{5}$	E''	$-\frac{12}{20}$	$\frac{6}{20}$	E_g	$-\frac{4}{5}$	$\frac{2}{5}$	E'_1	-1	$\frac{1}{2}$
$\sqrt{\frac{5}{4\pi}}\sqrt{3}\,zx$	2	1	πc												
$\sqrt{\frac{5}{4\pi}}\sqrt{3}\,xy$	3	2	δs	E_{2g}	$-\frac{2}{5}$	$-\frac{4}{5}$	E'	$\frac{21}{40}$	$\frac{12}{40}$	B_{2g}	$-\frac{4}{5}$	$\frac{12}{5}$	E'_2	$\frac{7}{8}$	$\frac{1}{2}$
$\sqrt{\frac{5}{4\pi}}\left(\sqrt{3}/2\right)(x^2-y^2)$	4	2	δc							B_{1g}	$\frac{11}{5}$	$-\frac{8}{5}$			

In deriving the reparameterization of the angular-overlap model for linearly ligating ligands, a nomenclature was used which emphasizes its resemblance to the conventional crystal-field model. If the parameters of this model are taken as freely adjustable, *i.e.* not calculable by electrostatics, the two models are indistinguishable. Therefore calculations performed within the crystal-field model can for linearly ligating ligands be translated into the angular-overlap model, and vice versa. The parameters I_L of *Yamatera* (20) are, when the point-charge model interpretation is neglected, then equal to the I^L reparameterization parameters; the same is true for the parameters $q\,G_L$ of *Ilse* and *Hartmann* (21) and $\mu\,B_L$ of *Ballhausen* (19), apart from certain constant factors from Slater orbitals.

A few examples will serve to illustrate how Table 8 can be translated into crystal-field results.

For the D_{3h} trigonal bipyramidal case of five ligands the d orbital energies of Table 8 are, in reparameterized form:

$$\overline{hda_1'} = \overline{hd\sigma} = \left[-\frac{3}{7}I^2 + \frac{9}{28}I^4\right] + \left[\frac{4}{7}I^2 + \frac{4}{7}I^4\right]$$

$$\overline{hde''} = \overline{hd\pi} = \left[-\frac{3}{14}I^2 - \frac{3}{14}I^4\right] + \left[\frac{2}{7}I^2 - \frac{8}{21}I^4\right] \qquad (52)$$

$$\overline{hde'} = \overline{hd\delta} = \left[\frac{3}{7}I^2 + \frac{3}{56}I^4\right] + \left[-\frac{4}{7}I^4 + \frac{2}{21}I^4\right]$$

where in each energy expression the first parenthesis contains the contribution from the three equatorial ligands and the second parenthesis that from the two axial ones. Putting for the equatorial parameters $I^L = \mu_1\,B_L$ and for the axial parameters $I^L = \mu_2 B_L$ the results of Eq. (52) are identical to those given by *Ballhausen* and *Jørgensen* (22). For the special choice $I^2 = 2\,I^4$, *Basolo* and *Pearson* (23, 24) have given the results of the crystal-field model for different symmetries in terms of the parameter Dq, where Dq is a single parameter, equal to $(1/10)\,(3e_\sigma' - 4e_\pi')$ of Eq. (26). The ratio $I^2/I^4 = 2$ corresponds, according to Eq. (37a), to $e_\sigma'/e_\pi' = 29/13$ which, combined with Eq. (26), gives $e_\sigma' = (58/7)\,Dq$ and $e_\pi' = (26/7)Dq$. Their Table 5 of (23) and Table 2.4 of (24) can be obtained directly by inserting these values for e_σ' and e_π' into Table 8.

6. Models, Parameters and Experiments

Ligand-field models serve the purpose of parameterizing experiments[9]. Their beauty and applicability stem from their derivation from the elementary theory of atomic spectra: they are first-order perturbation models based upon a basis set of l functions (assumption I, p. 71), and hence the interelectronic repulsion within the l shell may be accounted for in terms of *Condon* and *Shortley* parameters or *Racah* parameters. We obtain the expanded radial function model (*25, 13, 8*).

The first-order perturbation model upon an l basis without further assumptions may be called the non-additivity ligand-field model. This was discussed on p. (89) and briefly described in Eq. (42). With this model it is possible to assess by symmetry the number of one-electron parameters which are inherent in a given problem. For example, for d electrons in an octahedral field there is only one parameter [Eq. (26)]: the energy difference Δ.

A first-order perturbation model upon an l basis, with the additional assumption of the additivity of single-ligand perturbation contributions, may be called an additivity ligand-field model (*31, 16*). The angular-overlap model is such an additivity model, whether taken for the special case of linearly ligating ligands and reparameterized in the present paper, or for more general cases. We shall not discuss the general cases of AOM here but only refer to the discussion given in connection with Eq. (12) and that given at the end of Ref. (*10*).

The use of an l basis causes certain symmetry restrictions which do not arise from the symmetry of the perturbation itself.

Some special symmetry restrictions arise from the gerade inversion character of the basis. These have been discussed in terms of the concept of the holohedrized symmetry (*12, 27, 13, 8, 4*), which is a translation of the consequences of a symmetry property of the basis set (in this case its inversion symmetry) into an effective symmetry of the perturbation.

More generally, it is the detailed angular properties of the d functions which, combined with the symmetry of the surroundings, are responsible for the coefficients to the e_t or e_t' parameters of AOM. Understood in this way, Table 8 consists solely of symmetry coefficients (cf. footnote to p. 92). The reason is that, once the coordinates of one of the coordinating atoms (the ligators) are fixed, those of the rest are determined by

---

[9] The least restricted model [Ref. (*5*) p. 199] is a first-order perturbation model upon a basis which is not an l basis, but which transforms as an l basis under the molecular point-group symmetry. This model is satisfactory when only one l electron is present in the partially filled shell, but gives rise to a tremendous number of interelectronic repulsion parameters for l^n configurations with $n > 1$ [Ref. (*5*) p. 229]. In this case it is therefore of little practical use.

symmetry. There are no independent geometrical parameters in these cases when the symmetries are high and the number of ligators low.

It should, however, be emphasized that it is not a general feature of AOM that its coefficients to e_t and e_t' are solely symmetry-based. In general they will depend on geometrical parameters as well. Examples of this have been shown in Ref. (13), where it was stressed that the first prerequisite to make the model really useful would be a detailed geometrical knowledge of the chromophoric system to which the model is to be applied.

There are some other model degeneracies which are associated with a further restricted AOM for which only the e_σ parameters are assumed to be non-vanishing (28). In this case we obtain from Eq. ($\overline{37\,a}$)

$$e_\pi' = (3/7)I^2 - (5/21)I^4 = 0 \tag{53}$$

or

$$I^2/I^4 = 5/9 \tag{54}$$

as also mentioned by Kibler (29). For a single ligand, this condition gives a degeneracy of the four d orbitals which do not belong to the σ class. Similarly, for other chromophores, all the d orbitals which by symmetry cannot form σ bonds will remain degenerate. For a square complex of D_{4h} symmetry with four ligands, we obtain, for example,

$$hd\delta c = 3e_\sigma; hd\sigma = e_\sigma$$
$$hd\pi s = hd\pi c = hd\delta s = 0 \tag{55}$$

where we have chosen the unbarred operator which most clearly demonstrates the energy ratios and also the sum rule (9, 8): that the sum of the coefficients to e_t equals the number of ligands.

The comparison of AOM with experiment has two aspects: its value as a model for parameterizing experiments, and its value as a model whose parameters have a chemical significance.

The parameterization is easy to discuss for the angular-overlap model for linearly ligating ligands, since this is parametrically equivalent to the crystal-field model whose merits in this respect, for both d and f period complexes, are indisputable. The more general AOM has more parameters and will therefore be even more flexible for parameterization of experiments.

The chemical significance of the AOM parameters is much more difficult to discuss. The problem may be compared with that of establishing the chemical significance of nephelauxetism which quite analogously is measured by a parameter derived from experiments.

There is no elaborate theory that can account properly for the nephelauxetic parameter and the same is likely to be true for the AOM parameters. Still, nephelauxetism has proved a useful concept in inorganic chemistry and we believe the same will be true of the AOM parameters.

Let us draw attention first to the qualitative fact that Δ is positive [Eq. (26)] for all octahedral complexes and negative for tetrahedral ones. This is interpreted in terms of AOM parameters as

$$3e'_\sigma > 4e'_\pi \tag{56}$$

which is a satisfactory general property.

For the octahedron it is useful to define

$$\Delta_\sigma \equiv 3e'_\sigma; \; \Delta_\pi \equiv 4e'_\pi \tag{57}$$

so that, according to Eq. (26),

$$\Delta = \Delta_\sigma - \Delta_\pi \tag{58}$$

However, it should be emphasized that for octahedral complexes it is only this energy difference that can be determined and not the individual terms Δ_σ and Δ_π.

A quantitative comparison of AOM with experiment demands a situation where e'_σ and e'_π can be determined independently. This requires careful design of experiment to overcome the inherent problems.

Until now it has been done only for the case of chromium(III) doped in the mineral beryl (30). Here the chromium(III) ion is surrounded by six oxygen ions with a site symmetry D_3 for which there are two independent geometrical parameters which are approximately known from the crystal structure. The non-additivity model here tells us that there are three independent one-electron parameters (13) and these can, using AOM, be expressed in terms of the two parameters e'_σ and e'_π. From polarized crystal spectra the positions of four transitions are approximately known, allowing the determination of three one-electron parameters, plus the *Racah* parameter B.

The results are most satisfactory since the three parameters required by symmetry can be reproduced by the two AOM parameters. Using Eq. (57) the values found are

$$e'_\sigma = 10.5 \; kK; \; e'_\pi = 3.5 \; kK$$
$$\Delta_\sigma = 31.5 \; kK; \; \Delta_\pi = 10.5 \; kK \tag{59}$$

giving ratios $e'_\sigma/e'_\pi = 3$, and $\Delta_\sigma/\Delta_\pi = 9/4$, not unreasonable for a chromium(III)-oxygen bond (see p. 102).

It should be mentioned that these values have a large uncertainty stemming from the two main problems associated with this analysis: the extraction of the band positions from the polarized spectra and the extraction of the geometrical parameters for the doped chromium(III) ions from the crystal structure.

Other detailed comparisons (7, 31) of AOM with experiment have been done on tetragonal chromium(III) systems of the type $[Cr(NH_3)_4 L_2]$, also based upon the octahedron.

Here no geometrical parameters are required, and the non-additivity model tells us that again three one-electron parameters are inherent. This time, however, when AOM is used, these parameters are expressed in terms of the four parameters $e'_{\sigma N}$, $e'_{\sigma L}$, $e'_{\pi N}$, and $e'_{\pi L}$. Defining the non-additivity model parameters as

$$\Delta \equiv he(O_h) - ht_2(O_h)$$
$$\Delta(e) \equiv hb_1(D_{4h}) - ha_1(D_{4h}) = hd\delta c - hd\sigma \quad (60)$$
$$\Delta(t_2) \equiv hb_2(D_{4h}) - he(D_{4h})$$
$$= hd\delta s - hd\pi s = hd\delta s - hd\pi c$$

where

$$he(O_h) \equiv (1/2)[hb_1(D_{4h}) + ha_1(D_{4h})]$$
$$ht_2(O_h) \equiv (1/3)[hb_2(D_{4h}) + 2he(D_{4h})] \quad (61)$$

we obtain, using Eq. (57) and Table 8,

$$\Delta_{\sigma L} - \Delta_{\pi N} = \Delta - \frac{2}{3}\Delta(t_2) - \Delta(e)$$

$$\Delta_{\sigma N} - \Delta_{\pi N} = \Delta - \frac{2}{3}\Delta(t_2) + \frac{1}{2}\Delta(e) \quad (62)$$

$$\Delta_{\pi L} - \Delta_{\pi N} = -2\Delta(t_2)$$

so that, having derived Δ, $\Delta(e)$ and $\Delta(t_2)$ from experiment, $\Delta_{\sigma L}$, $\Delta_{\sigma N}$, and $\Delta_{\pi L}$ can be calculated relative[10] to $\Delta_{\pi N}$. Since ammonia

---

[10] It should be noted that Eq. (62) may be written

$$e'_{\sigma L} - \frac{4}{3}e'_{\pi N} = \frac{1}{3}\Delta - \frac{2}{9}\Delta(t_2) - \frac{1}{3}\Delta(e)$$

$$e'_{\sigma N} - \frac{4}{3}e'_{\pi N} = \frac{1}{3}\Delta - \frac{2}{9}\Delta(t_2) + \frac{1}{6}\Delta(e) \quad (63)$$

$$e'_{\pi L} - e'_{\pi N} = -\frac{1}{2}\Delta(t_2)$$

so that, if the e' parameters are used, the zero point for $e'_{\sigma L}$ and $e'_{\sigma N}$ becomes different from that of $e'_{\pi L}$.

has no π electrons available for bonding toward the central ion, $\Delta_{\pi N}$ would be expected to be quite small if the interpretation of the AOM parameters is physically adequate. Here it is worth mentioning that in all known cases $\Delta_{\pi L} - \Delta_{\pi N}$ has been found to be positive, or $\Delta_{\pi L} > \Delta_{\pi N}$. Further it seems to be a rule that the series of ligands corresponding to increasing Δ_σ values is the same as that corresponding to increasing Δ_π values. Ammonia and amines make up the only exceptions to this rule, which is in agreement with the fact that these ligands have no π electrons available.

We want finally to mention that for $L = OH^-$ (treated optimistically as a linearly ligating ligand) the results found were (7) $\Delta_{\sigma OH} - \Delta_{\pi N} = 27$ kK; $\Delta_{\pi OH} - \Delta_{\pi N} = 9$ kK so that, assuming $\Delta_{\pi N} = 0$, $\Delta_{\sigma OH}/\Delta_{\pi OH} = 3$. When this ratio is compared with that for the oxygen ion in beryl [Eq. (59)], one draws the chemically reasonable conclusion that the relative π antibonding effect upon the d orbitals is somewhat smaller with OH^- than with O^{--}.

With these tetragonal systems the main problems are the determination of the split components of the cubic parentage absorption bands and the fact that only three AOM parameters can be determined in each experiment. Here it helps a great deal that many complexes of this kind are available.

7. Conclusions

The angular-overlap model for d electrons, applied to systems with linearly ligating ligands, uses the single-ligand parameters

$$e'_\sigma \equiv hd\sigma - hd\delta$$

$$e'_\pi \equiv hd\pi - hd\delta \tag{64}$$

where e'_σ and e'_π are the energies of the $d\sigma$ orbital and the $d\pi$ orbitals, respectively, measured relative to that of the $d\delta$ orbitals. So the parameters are one-electron energy differences referring to a single linear central ion-to-ligand bond. As orbital energy differences, these parameters are the more physically interesting ones.

The alternative parameters I^2 and I^4 refer directly to an operator equivalent method (26), an irreducible tensor operator method (15) of handling the general first-order perturbation model.

However, in addition to being useful for calculation purposes (6), these parameters are applicable for translating experimental results which, in the case of linearly ligating ligands, have been analysed in terms of the electrostatic or crystal field model into the one-electron parameters of the angular-overlap model.

When the parameters e'_σ and e'_π have been determined by experiment, they represent one-electron energy differences. In the angular overlap model the e'_σ parameter is, apart from δ-bond contributions, interpreted as due to the d orbitals becoming σ-antibonding by combination with ligand σ orbitals. As such, e'_σ should always be positive. Similarly, e'_π is due to the d orbitals becoming π-antibonding, $e'_\pi > 0$, or π-bonding, $e'_\pi < 0$, by combination with filled or unfilled ligand orbitals, respectively. However, the formal identity between the crystal-field model and the reparameterized angular-overlap model causes an electrostatic contribution to the orbital energy differences to disturb the interpretation of e'_σ and e'_π, not as one-electron energy differences, but as expressions of the bonding effects mentioned.

The complete equivalence between the two parameterizations of AOM, conveyed by the one-to-one correspondence between their parameters [Eq. $(\overline{37}\,a)$], arises from the fact that they are both based on the first-order perturbation formalism, on the cylindrical symmetry of the central ion-to-ligand bond, and on the additivity of single-ligand perturbation contributions.

An important feature shown here is that it is necessary to assume additivity only for the non-spherical terms (8) in the perturbations. Since the spherical term, in so far as it can be defined, probably is the dominating one, this gives us a kind of explanation as to why the additivity assumption works so well in parameterizing experiments even though it is known that ligand-ligand overlap is not unimportant.

It has long been a theoretical puzzle (32, 33) that the crystal-field model works so well, even though its own interpretation of its parameters has been shown to be physically inadequate. The fact that the crystal-field model for the special case of linearly ligating ligands is parametrically equivalent to the angular-overlap model, whose parameters e'_σ and e'_π refer directly to the bonding process, may eventually solve this puzzle.

Appendix 1

Spherical Harmonic Analysis of Homogeneous Polynomia

A homogeneous polynomium f_l (x, y, z) may always be written

$$f_l = h_l + r^2 f_{l-2} = h_l + r^2 h_{l-2} + r^4 h_{l-4} + \ldots \qquad (65)$$

where h_l (x, y, z) is a solid spherical harmonic for which the equation

$$\nabla^2 h_l = 0 \tag{66}$$

is valid, and where in Eq. (65) the last term is h_1 for l odd and h_0 for l even. We shall call the expression of Eq. (65) an expansion of f_l into spherical harmonics, even though we emphasize [Ref. (15) p. 213] that in general

$$\nabla^2 f_l \neq 0 \tag{67}$$

and that ∇^2 acting on a solid harmonic h_l multiplied by a power of r, apart from a positive constant, only reduces that power by 2 units. The following formula was given by *Hobson* [Ref. (18) p. 120]

$$\nabla^2 r^q h_l = q[(2l + 1) + q] r^{q-2} h_l \tag{68}$$

It is often of interest to perform an expansion of the type of Eq. (65). *Hobson* has developed a projection operator which extracts (projects) the harmonic h_l out of the homogenous polynomial f_l. This operator $\mathfrak{S}$ is an irreducible tensorial operator of the degree zero for the three-dimensional rotation group and has the form [Ref. (18) p. 127)]

$$\mathfrak{S} = 1 - \{r^2/[2(2l - 1)]\}\nabla^2 + \{r^4/[2 \times 4 \times (2l - 1)(2l - 3)]\}\nabla^4 - \ldots \tag{69}$$

One then has

$$\mathfrak{S} f_l = h_l \tag{70}$$

and using Eqs. (65) and (66)

$$\nabla^2 f_l = \nabla^2 r^2 h_{l-2} + \nabla^2 r^4 h_{l-4} + \ldots \tag{71}$$

where, according to Eq. (68), the first term is the harmonic of the degree $(l-2)$, which can be projected out using the operator $\mathfrak{S}$. In this way by alternating use of $\mathfrak{S}$ and ∇^2, and by application of Eq. (68), the expansion of Eq. (65) can be performed. Let us take as an example

$$f_2 = x^2 = h_2 + r^2 h_0 \tag{72}$$

in which case only the first two terms of $\mathfrak{S}$ contribute. One obtains

$$\mathfrak{S}x^2 = x^2 - (1/6)(x^2 + y^2 + z^2) 2 = (2/3)x^2 - (1/3)y^2 - (1/3)z^2 = h_2 \tag{73}$$

which, apart from a constant of normalization, may be recognized as the $d\sigma$ function about the X axis. Using now Eq. (72) and Eq. (68) for $q = 2$ and $l = 0$, one obtains

$$\nabla^2 x^2 = 6h_0 = 2 \tag{74}$$

so that

$$x^2 = (1/3) [2x^2 - y^2 - z^2] + (1/3) [x^2 + y^2 + z^2] \qquad (75)$$

Alternative methods of performing the expansion of Eq. (65) have been described in [Ref. (18) p. 148].

We have seen how the operator $\mathfrak{S}$ is able to project h_l out of f_l. There is a general alternative manner of obtaining h_l which belongs to the irreducible tensorial set of the degree l of the three-dimensional rotation group. This is another irreducible tensorial operator of the degree zero, the operator for the resolution of the identity within l space [Ref. (11) p. 383] which may be written

$$\mathfrak{J}^l = \sum_t O_t^{(l)} \qquad (76)$$

where $O_t^{(l)}$ is the standard projection operator, eq. (82).

Sometimes there is a special and easier manner of obtaining h_l. We exemplify this by the totally symmetric cubic harmonic of the degree 4. This harmonic must consist of a linear combination of $x^4 + y^4 + z^4$ (of cubic symmetry) and r^4 (of spherical symmetry, and therefore also of cubic symmetry), since only two linearly independent homogeneous polynomia simultaneously are of cubic symmetry and of the degree 4. This means that the spherical harmonic must have the form

$$h_4(a_1) = x^4 + y^4 + z^4 + c\, r^4 \qquad (77)$$

with the condition, $\nabla^2 h_4(a_1) = 0$, which, using Eq. (77), gives $c = -3/5$. The factor n required for normalizing $h_4(a_1)$ to $4\pi/(2L+1) = 4\pi/9$ may be found by using the unification operator (15)

$$\mathfrak{N}_t^l = 2^l\, l!\, [(2l)!]^{-1} \mathfrak{D}_t^l = [1 \times 3 \times 5 \times \ldots \times (2l-1)]^{-1} \mathfrak{D}_t^l \qquad (78)$$

for which the relation

$$\mathfrak{N}_t^l \, \mathfrak{H}_{t'}^{l'} = \delta(l, l')\, \delta(t, t') \qquad (79)$$

is valid. In Eq. (78) the differential operator $\mathfrak{D}_t^l$ obtains when the substitutions (6) $x \to \partial/\partial x$, $y \to \partial/\partial y$, $z \to \partial/\partial z$ are made in $\mathfrak{H}_t^l$ [normalized to $4\pi/(2l+1)$ over the unit sphere, (15) p. 213]. So one obtains

$$\frac{1}{3 \times 5 \times 7}\, n \left[\frac{\partial^4}{\partial x^4} + \frac{\partial^4}{\partial y^4} + \frac{\partial^4}{\partial z^4} - \frac{3}{5}\, \nabla^4 \right] n \left[x^4 + y^4 + z^4 - \frac{3}{5} r^4 \right] = 1 \qquad (80)$$

105

C. E. Schäffer

giving the value $(5/12) \sqrt{21}$ for the normalization constant n, so that

$$\mathfrak{H}_{a_1}^4 = (5/12) \sqrt{21} \, [x^4 + y^4 + z^4 - (3/5) \, r^4]$$
$$= (1/6) \sqrt{21} \, [x^4 + y^4 + z^4 - 3\,y^2\,z^2 - 3\,z^2\,x^2 - 3\,x^2\,y^2] \qquad (81)$$

The operator $\mathfrak{S}$ of Eq. (69) and $\mathfrak{J}$ of Eq. (76) projects h_l out of f_l. However, often one is more interested in obtaining the contents of irreducible tensorial components $\mathfrak{H}_t^l$ in f_l. This can be projected out of f_l using the idempotent [Ref. (15) p. 212]

$$O_t^{(l)} = \mathfrak{H}_t^l \, \mathfrak{N}_t^l \qquad (82)$$

We illustrate this by returning to our example of $f_2 = x^2$. The only non-vanishing $\mathfrak{H}_t^2$ occurs for $t = \sigma$ and δc. We find using $O_\sigma^{(l)}$

$$(1/3) \, \mathfrak{H}_\sigma^2 \, [- (1/2) \, \partial^2/\partial x^2] \, x^2 = - (1/3) \, \mathfrak{H}_\sigma^2 \qquad (83)$$

and using $O_{\delta c}^{(l)}$

$$(1/3) \, \mathfrak{H}_{\delta c}^2 \, [(\sqrt{3}/2) \, \partial^2/\partial x^2] \, x^2 = (\sqrt{3}/3) \, \mathfrak{H}_{\delta c}^2 \qquad (84)$$

and on adding the results of Eqs. (83) and (84), using Table 7, one obtains h_2 as given in Eq. (75).

It is sometimes of interest to consider a special type of homogeneous polynomium which is the product of two harmonics

$$f_l = f_{m+n} = h_m \, h_n \qquad (85)$$

Repeated application of ∇^2 on such polynomia gives particularly simple expressions of which we state the first few, which may be readily extrapolated

$$\nabla^2 \, [h_m \, h_n] = 2 \sum_p \frac{\partial h_m}{\partial p} \frac{\partial h_n}{\partial p}$$

$$\nabla^4 \, [h_m \, h_n] = 2^2 \sum_p \sum_q \frac{\partial^2 h_m}{\partial p \, \partial q} \frac{\partial^2 h_n}{\partial p \, \partial q} \qquad (86)$$

$$\nabla^6 \, [h_m \, h_n] = 2^3 \sum_p \sum_q \sum_s \frac{\partial^3 h_m}{\partial p \, \partial q \, \partial s} \frac{\partial^3 h_n}{\partial p \, \partial q \, \partial s}$$

where in Eq. (86) p, q, and s independently take on the values x, y, and z.

For $f_4 = x^2 y^2$, for example, the choice must be $h_m = h_n = xy$ and we get directly

$$\nabla^4 x^2 y^2 = \nabla^2 (2 x^2 + 2 y^2) = 8 \tag{87}$$

or, by using Eq. (86)

$$\nabla^4 (xy) (xy) = 2^2 (1 + 1) = 8 \tag{88}$$

where $(p, q) = (x, y)$ and $(p, q) = (y, x)$ make up two different choices in the summation.

Appendix 2

Evaluation of Even 3-l Symbols through Integration by Differentiation

By successive application of Eq. (68) this may be generalized to the following expression

$$\nabla^{2n} r^{2n} h_{l-2n} = \frac{2n + 1}{1} \times \frac{2n + 3}{3} \times \cdots \frac{2n + 2(l - 2n) + 1}{2(l - 2n) + 1}$$
$$\times (2n)! \, h_{l-2n} \tag{89}$$

Letting now ∇^{2n} act on f_l, we obtain

$$\nabla^{2n} f_l = \nabla^{2n} r^{2n} h_{l-2n} + r^2 g_{l-2n-2} \tag{90}$$

where ∇^{2n} has annihilated the first n terms of Eq. (65), and where the first term on the right-hand side of Eq. (90) is a spherical harmonic of the degree $(l - 2n)$ [Eq. (68)] and the second term does not even contain harmonics of this degree. From Eq. (90) one obtains, using Eq. (89),

$$\nabla^{2n} r^{2n} h_{l-2n} = \nabla^{2n} f_l - r^2 g_{l-2n-2}$$
$$= \frac{(2n + 1) (2n + 3) \times \cdots \times [2n + 2(l - 2n) + 1]}{1 \times 3 \times 5 \cdots \times [2(l - 2n) + 1]} (2n)! \, h_{l-2n} \tag{91}$$

or

$$h_{l-2n} = \frac{1 \times 3 \times 5 \times \cdots \times [2(l - 2n) + 1]}{(2n + 1) (2n + 3) \times \cdots \times [2n + 2(l - 2n) + 1]} \frac{1}{(2n)!}$$
$$[\nabla^{2n} f_l - r^2 g_{l-2n-2}] \tag{92}$$

The content of $\mathfrak{H}_t^{l-2n}$ in h_{l-2n} may now be obtained by operating on the right-hand side of Eq. (92) with [Eqs. (82) and (78)]

$$O_t^{(l-2n)} = \{1 \times 3 \times 5 \times \ldots \times [2(l-2n)-1]\}^{-1} \, \mathfrak{H}_t^{l-2n} \, \mathfrak{D}_t^{l-2n} \quad (93)$$

This operator annihilates $r^2 \, g_{l-2n-2}$ and the final result is therefore

$$[2(l-2n)+1] \, \{(2n+1)(2n+3) \times \ldots \times [2n+2(l-2n)+1](2n)!\}^{-1}$$
$$\times \, \mathfrak{H}_t^{l-2n} \, \mathfrak{D}_t^{l-2n} \, \nabla^{2n} \, f_l \quad (94)$$
$$= \frac{(l-n)! \, 2^{l-2n} \, [2(l-2n)+1]}{n! \, [2n+2(l-2n)+1]!} \, \mathfrak{H}_t^{l-2n} \, \mathfrak{D}_t^{l-2n} \, \nabla^{2n} \, f_l$$

We are now able to perform an integration over the unit sphere of the product of three solid harmonics $\mathfrak{H}_{t_1}^{l_1}$, $\mathfrak{H}_{t_2}^{l_2}$, and $\mathfrak{H}_{t_3}^{l_3}$ with $l_1+l_2+l_3$ even. For this integral the following relation is valid [Ref. (15), Eq. (76)]:

$$(1/4\pi) \int \mathfrak{H}_{t_1}^{l_1} \, \mathfrak{H}_{t_2}^{l_2} \, \mathfrak{H}_{t_3}^{l_3} \, dS = \begin{pmatrix} l_1 & l_2 & l_3 \\ \sigma & \sigma & \sigma \end{pmatrix} \begin{pmatrix} l_1 & l_2 & l_3 \\ t_1 & t_2 & t_3 \end{pmatrix} \quad (95)$$

The integral itself may be evaluated by using the operator of Eq. (94) on, for example, the product $\mathfrak{H}_{t_1}^{l_1} \, \mathfrak{H}_{t_2}^{l_2}$ to project out of this product its content of $\mathfrak{H}_{t_3}^{l_3}$. Remembering that $\mathfrak{H}_{t_2}^{l_2}$ is normalized to $4\pi/(2l_2+1)$, one obtains

$$(1/4\pi) \int \mathfrak{H}_{t_1}^{l_1} \, \mathfrak{H}_{t_2}^{l_2} \, \mathfrak{H}_{t_3}^{l_3} \, dS = \frac{[(l_1+l_2+l_3)/2]! \, 2^{l_1}}{[(l_3+l_1-l_2)/2]! \, [l_1+l_2+l_3+1]!}$$
$$\mathfrak{D}_{t_3}^{l_3} \, \nabla^{l_3+l_1-l_2} \, [\mathfrak{H}_{t_1}^{l_1} \, \mathfrak{H}_{t_2}^{l_2}] \quad (96)$$

Using now Eqs. (96) and (95) together with the expression for $\begin{pmatrix} l_1 & l_2 & l_3 \\ \sigma & \sigma & \sigma \end{pmatrix}$ of Eq. (31), one obtains

$$\begin{pmatrix} l_1 & l_2 & l_3 \\ t_1 & t_2 & t_3 \end{pmatrix} = 2^{l_1} \, [(l_2+l_3-l_1)! \, (l_3+l_1-l_2)! \, (l_1+l_2-l_3)!$$
$$\times (l_1+l_2+l_3+1)!]^{-\frac{1}{2}} \, [(l_2+l_3-l_1)/2]! \, [(l_1+l_2-l_3)/2]! \quad (97)$$
$$\mathfrak{D}_{t_3}^{l_3} \, \nabla^{l_1+l_2-l_3} \, [\mathfrak{H}_{t_1}^{l_1} \, \mathfrak{H}_{t_2}^{l_2}]$$

In Eqs. (96) and (97) l_2 appears to play a unique role. However, the l's may of course be permuted in the formulae, provided the exponent to ∇ remains non-negative.

In calculating the integrals which are of interest for the reparameterized angular overlap model, we put $l_2 = L$ and $l_1 = l_3 = l$ and at the same time we renormalize the l functions to unity. Thereby Eq. (96) becomes

$$<lu|\mathfrak{C}_T^L|lv> = \frac{(2l+1)\,[(2l+L)/2]!\,2^L}{[(2l-L)/2]![2l+L+1]!}$$

$$\mathfrak{D}_T^L \,\nabla^{2l-L}\,[\mathfrak{H}_u^l \,\mathfrak{H}_v^l] \tag{98}$$

where in integrations over the unit sphere surface harmonics $\mathfrak{C}_T^L$ and solid harmonics

$$\mathfrak{H}_T^L = r^L \,\mathfrak{C}_T^L \tag{99}$$

are synonymous [Ref. (15) p. 213]. Eq. (97) similarly becomes

$$\begin{pmatrix} l & L & l \\ u & T & v \end{pmatrix} = 2^L\,[(L!)^2\,(2l-L)!\,(2l+L+1)!]^{-\frac{1}{2}}$$

$$\times\,[(L/2)!]^2\,\mathfrak{D}_T^L \,\nabla^{2l-L}\,[\mathfrak{H}_u^l \,\mathfrak{H}_v^l] \tag{100}$$

As an example of the application of Eq. (100) we take $L = l = 2$, $T = \pi s$, $u = \pi c$, and $v = \delta s$ to obtain

$$\begin{pmatrix} d & d & d \\ \pi c & \pi s & \delta s \end{pmatrix} = \begin{pmatrix} 2 & 2 & 2 \\ 2 & 1 & 3 \end{pmatrix} = 2^2\,[2^2 \times 2 \times 7!]^{-\frac{1}{2}} \tag{101}$$

$$\times\,\sqrt{3}\,\partial^2/\partial y\partial z\,\nabla^2\,[\sqrt{3}\,zx \times \sqrt{3}\,xy] = \sqrt{3/70}$$

in agreement with Table 6.

References

1. *Jørgensen, C. K.*: Quantum Theory of Atoms, Molecules, Solid State, p. 307; ed. *P. O. Löwdin*. New York: Academic Press 1966.
2. — Struct. Bonding *1*, 233 (1966).
3. — Chimia *23*, 292 (1969); *25*, 109 (1971).
4. — Modern Aspects of Ligand Field Theory. Amsterdam: North-Holland 1971.
5. *Griffith, J. S.*: The Theory of Transition Metal Ions. Cambridge: Univ. Press 1961.
6. *Harnung, S. E., Schäffer, C. E.*: Struct. Bonding *12*, 257 (1972).
7. *Glerup, J., Schäffer, C. E.*: Proc. XI ICCC Progress in Coordination Chemistry, p. 500; ed. *M. Cais*. Amsterdam: Elsevier 1968.
8. *Schäffer, C. E.*: Pure Appl. Chem. *24*, 361 (1970).

9. — *Jørgensen, C. K.:* Mol. Phys. *9*, 401 (1965).
10. — Struct. Bonding *5*, 68 (1968).
11. — Intern. J. Quantum Chem. *5*, 379 (1971).
12. — *Jørgensen, C. K.:* Mat.-fys. Medd. Selsk. *34* No. 13, (1965).
13. — Proc. Roy. Soc. (London) *A 297*, 96 (1967).
14. *Jørgensen, C. K.:* Struct. Bonding *1*, 3 (1966).
15. *Harnung, S. E., Schäffer, C. E.:* Struct. Bonding *12*, 201 (1972).
16. *Bradbury, M. I., Newman, D. J.:* Chem. Phys. Letters *1*, 44 (1967).
17. *Bacon, G. E., Gardner, W. E.:* Proc. Roy. Soc. (London) *A 246*, 78 (1966).
18. *Hobson, E. W.:* The Theory of Spherical and Ellipsoidal Harmonics. Cambridge: Univ. Press 1931.
19. *Ballhausen, C. J.:* Math.-fys. Medd. Selsk *29*, No. 4 (1954).
20. *Yamatera, H.:* Bull. Chem. Soc. Japan *31*, 95 (1958).
21. *Ilse, F. E., Hartmann, H.:* Z. Physik. Chem. *197*, 239 (1951).
22. *Ballhausen, C. J., Jørgensen, C. K.:* Math.-fys. Medd. Selsk. *29*, No. 14 (1955).
23. *Basolo, F., Pearson, R. G.:* Mechanisms of Inorganic Reactions. New York: Wiley 1958.
24. — — Mechanisms of Inorganic Reactions, 2. Edit. New York: Wiley 1967.
25. *Jørgensen, C. K.:* Discussions Faraday Soc. *26*, 110 (1958).
26. *Griffith, J. S.:* J. Chem. Phys. *41*, 576 (1964).
27. *Jørgensen, C. K.:* J. Phys. Radium *26*, 825 (1965).
28. — *Pappalardo, R., Schmidtke, H.-H.:* J. Chem. Phys. *39*, 1422 (1963).
29. *Kibler, M. R.:* Chem. Phys. Letters *8*, 142 (1971).
30. *Maegaard, H., Schäffer, C. E.:* to be published.
31. *Schäffer, C. E.:* Wave Mechanics — the First Fifty Years Eds. S. S. Chissick, W. C. Price, T. Ravensdale: London: Butterworths 1973.
32. *Freeman, A. J., Watson, R. E.:* Phys. Rev. *120*, 1254 (1960).
33. *Jørgensen, C. K.:* Absorption Spectra and Chemical Bonding in Complexes. Oxford: Pergamon 1962.

Received July 25, 1972

Salzebullioskopie

III Theorie und Anwendung zur Bestimmung des Kondensationsgrades polynuklearer Metallkomplexe *

B. Magyar

Laboratorium für Anorganische Chemie, Eidg. Techn. Hochschule, Zürich, Schweiz

Inhalt

---

* Auszug aus (22), zweite Mitteilung (4).

1. Einleitung

Zur Kenntnis des ionischen Aufbaus eines Komplexsalzes der allgemeinen Formel,

$$[(MB_bC_c\ldots)_n]G_{n\cdot g}\,, \tag{1.1}$$

wobei M das Zentralatom, B, C, ... irgendwelche Elemente der Liganden und G das Gegenion bedeuten, muß neben den Atomverhältnissen $1, b, c, \ldots$, und g, die mit quantitativen analytischen Methoden bestimmt werden, auch noch die Zahl n gefunden werden. Sie wird die Nuclearität des Komplexsalzes genannt. Der reziproke Wert von n ist also die Anzahl $(v = 1/n)$ freibeweglicher Partikeln, die pro Metall M beim Lösen des Komplexsalzes neben den g Gegenionen auftreten. Sind sowohl $b, c, \ldots$, g als auch n bekannt, so kann das Ionengewicht IG des Komplexions angegeben werden:

$$IG = n \cdot (A_M + b \cdot A_B + c \cdot A_C + \ldots)\,, \tag{1.2}$$

wobei A_B, A_C, ... die Atomgewichte der Elemente B, C, ... bedeuten. Daher wird bei der Ermittlung von n gelegentlich von der „Bestimmung von Ionengewichten" gesprochen. Im Prinzip könnte n auch kleiner als 1 sein, d.h. $v > 1$, was Zerfall des Komplexions anzeigt.

In der Chemie der Metallkomplexe kommt also der Bestimmung von v die gleiche Bedeutung zu wie der Bestimmung des Molekulargewichtes bei Stoffen, welche Molekeln zu Bausteinen haben. Als Möglichkeiten zur Bestimmung von Molekulargewichten und Ionengewichten bieten sich prinzipiell die Grenzgesetze verdünnter Lösungen an. Für niedermolekulare Stoffe werden meist die Kryoskopie und Ebullioskopie herangezogen, die im Falle von Komplexsalzen zu Gleichungen vom Typ (1.3) führen.

$$\frac{\Delta T}{K \cdot m} = v(m) = (g + 1/n) \cdot f \tag{1.3}$$

ΔT : die Schmelzpunktserniedrigung bzw. Siedepunktserhöhung
K : die kryoskopische bzw. ebullioskopische Konstante
m : die Molalität des Zentralatoms M (g Atom M pro kg H_2O)
$v(m)$: die scheinbare Teilchenzahl
f : praktischer osmotischer Koeffizient des Lösungsmittels

Da der osmotische Koeffizient f nicht zugänglich ist, müssen die gemessenen v-Werte auf die Konzentration $m = 0$ (wo $f = 1$) extrapoliert werden. Natürlich ist es unmöglich, Messungen bei beliebig kleinen

Konzentrationen des Fremdstoffes auszuführen, so daß ν bei $m = 0$ nur durch Extrapolation erhalten werden kann, was deshalb Schwierigkeiten macht, weil die Funktion nicht linear ist (s. Kurve A in Abb. 1.1). Um $1/n$ zu erhalten, muß zudem noch die Anzahl g der Gegenionen pro M vom Ordinatenabschnitt subtrahiert werden. Als Unterschied von zwei etwa gleich großen Zahlen wird deshalb $1/n$ mit großem Fehler behaftet sein. Kryoskopische und ebullioskopische Messungen an Salzen im Wasser als Lösungsmittel gestatten demzufolge kaum je eine eindeutige Aussage über die Nuclearität n.

Eine große Verbesserung brachte die Verwendung von Salz-Hydratschmelzen und eutektischen Mischungen von Eis mit gesättigter Inertsalzlösung als kryoskopisches Medium.

Salzkryoskopische Messungen, mit schmelzendem Glaubersalz ($Na_2SO_4 \cdot 10\,H_2O$) als Lösungsmittel, hat erstmals 1895 *Löwenherz* (1) ausgeführt. Die Salzkryoskopie zeigt gegenüber der normalen Kryoskopie zwei entscheidende Vorteile:

a) Die experimentell bestimmten scheinbaren Teilchenzahlen weisen nun eine lineare Abhängigkeit von der Konzentration des zu untersuchenden Stoffes (Fremdstoff) auf.

b) Solche Ionen, welche in der Hydratschmelze oder in der eutektischen Mischung bereits enthalten sind, verursachen keine Depression des Fixpunktes.

Als Beispiel können die mit Uranylsulfat erhaltenen Resultate dienen (Abb. 1.1). Kurve A illustriert die klassische Kryoskopie in Wasser und ist zu vergleichen mit der Geraden B, erhalten im $Na_2SO_4 \cdot 10\,H_2O$-Eis-Eutektikum. Es ist klar, daß man die Kurve A kaum zuverlässig auf $m = 0$ extrapolieren kann. Weiter erhält man durch die leicht zu bewerkstelligende Extrapolation von B direkt den reziproken Wert der Nuclearität $1/n = 1$, weil das Gegenion SO_4^{2-} nicht kryoskopisch wirksam ist. Demgegenüber ist vom extrapolierten Wert der Kurve A noch die Zahl g abzuziehen, da in Wasser als Medium auch das Gegenion natürlich zur Schmelzpunktserniedrigung beiträgt.

Während die Salzkryoskopie schon vor **75** Jahren eingeführt wurde, gibt es noch kein ebullioskopisches Analogon, bei dem als Medium eine Salzlösung verwendet wird. Eine solche „Salzebullioskopie" muß gegenüber der normalen Ebullioskopie dieselben Vorteile aufweisen wie die Salzkryoskopie gegenüber der üblichen Kryoskopie. Daß man an diese Möglichkeit bisher nicht dachte, hängt sicherlich damit zusammen, daß salzebullioskopische Messungen nicht einfach analog wie normale ebullioskopische Molekulargewichtsbestimmungen ausgeführt werden können, während zwischen Salzkryoskopie und klassischer Kryoskopie keine methodischen Unterschiede bestehen. Bei der klassischen Ebullioskopie

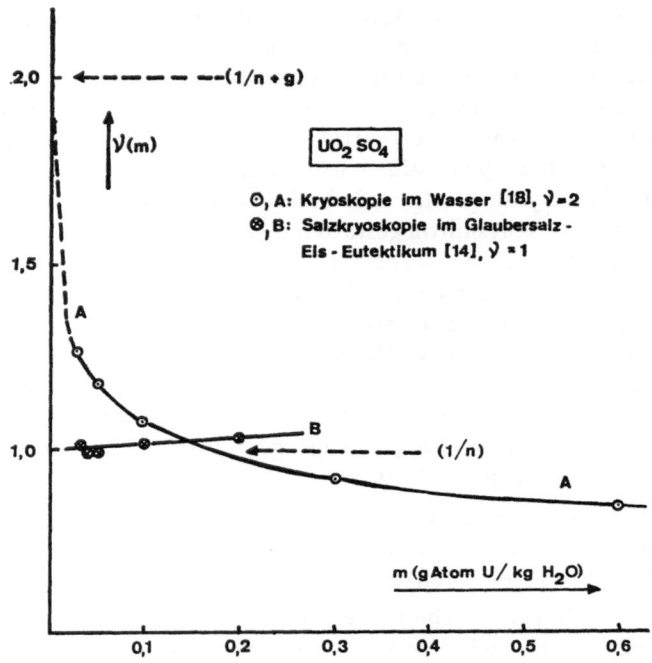

Abb. 1.1, Änderung der scheinbaren Anzahl $v(m)$ von Fremdpartikeln mit der Konzentration m des Fremdstoffes (Uranylsulfat). Die Literatur-Werte für die Depression des Fixpunktes ΔT wurde gemäß Gl. (1.3) in $v(m)$-Werte umgewandelt, wobei für K 1,86 (Kurve A) bzw. 1,78 (Kurve B) eingesetzt wurde

vergleicht man den Kochpunkt einer Lösung mit dem Kondensationspunkt des entstandenen Dampfes, denn letzterer ist beim Einkomponenten-Medium mit dem Siedepunkt des Lösungsmittels identisch. Diese Methode hat sich darum gut bewährt, weil sie gewährleistet, daß beide Temperaturen gleichzeitig, und was noch wichtiger ist, unter gleichem Druck gemessen werden, dessen Absolutwert unwesentlich ist. Eine Veränderung des äußeren Druckes δP verursacht nämlich — bei unveränderter Konzentration m — eine Änderung der Siedepunktserhöhung δT, welche folgender Beziehung gehorcht:

$$\left[\frac{\delta P}{\delta T}\right]_m = \frac{P^\circ}{55{,}51 \cdot E} \tag{1.4}$$

wobei E die ebullioskopische Konstante und P° der Dampfdruck des reinen Lösungsmittels bedeuten. Bei der klassischen Methode sorgt der Außendruck dafür, daß über Lösung und Lösungsmittel der gleiche Dampfdruck herrscht.

Anders ist es aber, wenn das Medium eine gesättigte Salzlösung ist, bei welcher durch Kondensation des Dampfes reines Wasser entsteht, also nicht das ebullioskopische Medium. Prinzipiell besteht die Möglichkeit, unter gleichem Außendruck die Siedepunktsdifferenz zweier gesättigter Inertsalzlösungen, mit festem Inertsalz als Bodenkörper, von denen die eine noch etwas Fremdsalz enthält, zu messen. Die praktische Ausführung scheitert jedoch daran, daß die Siedetemperatur von Lösungen mit festem Bodensatz wegen Siedeverzügen sehr schwierig genau zu bestimmen ist.

Der Zusatz eines Fremdsalzes zu einem Medium Inertsalz/Wasser erniedrigt natürlich dessen Dampfdruck nicht erst am Siedepunkt, sondern bei beliebiger Temperatur. Die Dampfdruckerniedrigung ΔP einer gesättigten Inertsalzlösung mit festem Inertsalz als Bodenkörper, welche bei Zugabe eines Fremdstoffes in Bezug auf die reine gesättigte Inertsalzlösung eintritt, kann nach der isopiestischen Methode gemessen werden, wie in dieser Arbeit erstmals gezeigt wird. Dieses ΔP kann ebenfalls zur Ermittlung der Anzahl ν_0 der Fremddionen pro Formel Fremdsalz herangezogen werden. Daß dabei thermodynamische Gleichgewichte gemessen werden, macht diese Methode für theoretische Studien besonders attraktiv. Ihre Ausführung ist jedoch sehr langwierig. Für serienmäßige Ionengewichtsbestimmungen muß man über eine schnellere Technik verfügen.

Gemäß der allgemein gültigen Beziehung 1.4 entspricht die Dampfdruckerniedrigung ΔP des Mediums, verursacht durch die Fremdsalzkonzentration m, einer isobaren Temperaturerhöhung ΔT, welche sich dann einstellt, wenn über beiden Medium-Proben (mit und ohne Fremdsalz) der gleiche Dampfdruck herrscht. In der ersten Veröffentlichung (2) über Salzebullioskopie wurde gezeigt, daß sich die isobare Temperaturerhöhung ΔT in Form eines stationären Temperaturunterschiedes ΔT_s messen läßt mit Hilfe der sog. Hill'schen Methode (3). Diese Methode besteht darin, daß man kleine Proben des Mediums (Inertsalz und Lösung) mit und ohne Fremdstoff in eine Atmosphäre bringt, die den Wasserdampfdruck des Mediums besitzt. Es stellt sich dann eine über längere Zeit konstant bleibende Temperaturdifferenz ΔT_s zwischen den beiden Proben ein, aus der auf die Zahl ν_0 der freibeweglichen Fremddionen geschlossen werden kann. Wir haben uns nun für den funktionellen Zusammenhang zwischen diesen Größen zu interessieren.

Wie alle Grenzgesetze verdünnter Lösungen, kann auch das salzebullioskopische Grenzgesetz nur für den Fall unendlicher Verdünnung streng abgeleitet werden (4). Zur Anwendung auf endliche Konzentrationen müssen außerthermodynamische Annahmen gemacht werden, deren Richtigkeit experimentell zu beweisen ist. Deshalb wurden viele Messungen an Salzen mit bekanntem ionischen Aufbau ausgeführt.

2. Theoretischer Teil

2.1. Das salzebullioskopische Grenzgesetz

Ein salzebullioskopisches System besteht aus 3 Komponenten mit 3 Phasen:

Komponenten: 1. H_2O Phasen: flüssig: kein Index
            2. Fremdsalz   festes Inertsalz: Index ('')
            3. Inertsalz   Gas (H_2O): Index (')

Nach der Gibbs'schen Phasenregel besitzt dieses System 2 Freiheitsgrade. Hält man den Partialdruck des Wassers konstant, so besitzt das System nun mehr einen einzigen Freiheitsgrad. Wählt man die Molalität der Komponente 2 (Fremdsalz) als freie Variable, so ist die Temperatur des Systems eindeutig festgelegt. Im System ohne Fremdstoff (Vergleichssystem) gibt es dagegen bei konstantem Partialdruck des Wassers keinen Freiheitsgrad mehr. Es soll nun die Temperaturdifferenz dT zwischen beiden Systemen (mit und ohne Fremdstoff) untersucht werden, die eintreten muß, wenn über beiden Lösungen der gleiche Partialdruck des Wassers herrscht.

Nach den Gesetzen der Thermodynamik besteht dann ein Gleichgewicht in einem Mehrkomponentensystem, wenn das chemische Potential der Komponenten in allen Phasen gleich ist. Dieses Gesetz kann hier folgendermaßen formuliert werden:

$$\frac{dT}{dm_2}\left\{ m_1 \frac{\delta \ln a_1'}{\delta T} - m_1 \frac{\delta \ln a_1'}{\delta T} + m_3 \frac{\delta \ln a_3''}{\delta T} - m_3 \frac{\delta \ln a_3}{\delta T} \right\} =$$

$$= m_2 \frac{\delta \ln a_2}{\delta m_2} + m_2 \frac{dm_3}{dm_2} \frac{\delta \ln a_2}{\delta m_3}, \qquad (2.1)$$

worin das Symbol (d) das Totaldifferenzial, (δ) die partielle Ableitung bezeichnet. m_1, m_2 und m_3 bedeuten die Anzahl Mole der Komponenten 1, 2 und 3 per 1 kg Wasser (= Molalität). Mit (a) ist die Aktivität bezeichnet, welche willkürlich auf die reine gesättigte Inertsalzlösung (= Medium) bezogen wird.

Bei der Ableitung dieser Gleichung wurde das chemische Potential üblicherweise in einen temperatur- und druckabhängigen und einen konzentrationsabhängigen Teil aufgespalten. Weiterhin wurden die Aktivitätsänderungen des Wassers und des Inertsalzes mit Hilfe der Gibbs-Duhem-Gleichung durch die Aktivitätsänderung des Fremdstoffes substituiert.

Die Temperaturabhängigkeit der Aktivitäten ist durch die folgende Gleichung gegeben:

$$\left[\frac{\delta \ln a_i}{\delta T}\right]_p = -\frac{\overline{H}_i - \overline{H}_i^0}{RT^2}, \ i = 1 \ \text{bzw.} \ 3 \qquad (2.2)$$

Hier bedeuten: $\overline{H}_i$ die partiale molare Enthalpie der Komponente i in der flüssigen Mischphase (Medium und Fremdsalz), und $\overline{H}_i^0$ die partiale molare Enthalpie in der reinen gesättigten Inertsalzlösung (Medium). Der Klammerausdruck auf der linken Seite der Gl. (2.1) nimmt damit den folgenden Wert an:

$$m_1 \frac{\overline{H}_1 - H_1'}{RT^2} + m_3 \frac{\overline{H}_3 - H_3'}{RT^2} = \frac{W}{RT^2} = 1/E \qquad (2.3)$$

$(H_1' - \overline{H}_1)$ bedeutet die Änderung der molaren Enthalpie des Wassers beim Übergang aus der Lösung in die Dampfphase (Verdampfungswärme). $(H_3' - \overline{H}_3)$ stellt die Kristallisationswärme des Inertsalzes (= differentielle Lösungswärme bei der Sättigungskonzentration) dar. Verdampfen von 1 kg Wasser aus diesem System zieht das Auskristallisieren von m_3 Formelgewichten Inertsalz nach sich. Im Grenzfall, wo $m_2 = 0$, ist $m_3 = m_3^0$, nämlich gleich der Löslichkeit des Inertsalzes in 1 kg Wasser. Die beiden mit der entsprechenden Molzahl multiplizierten Enthalpiedifferenzen bedeuten also zusammen die Verdampfungswärme für 1 kg Wasser aus gesättigter Inertsalzlösung, wenn $m_2 \to 0$. Diese wurde in Gl. (2.3) mit W bezeichnet, (= Enthalpieänderung einer Mediummenge, welche 1 kg Wasser enthält). Wenn man nur verdünnte Lösungen untersucht, so ist ΔT klein, so daß W/RT^2 eine Konstante bleibt. Der reziproke Wert dieser Größe soll als salzebullioskopische Konstante E bezeichnet werden.

Die Gl. (2.1) ist thermodynamisch streng gültig. Sie kann aber erst dann als Grundlage zur Ableitung eines Grenzgesetzes dienen, wenn die Aktivität a_2 mit Hilfe eines willkürlichen Ansatzes in Zusammenhang mit der unabhängigen Variablen m_2 gebracht wird. Die Notwendigkeit des Heranziehens außer-thermodynamischer Annahmen über die Konzentrationsabhängigkeit der Aktivitäten zur Ableitung der Grenzgesetze wurde von *Haase* (5) sehr eindrucksvoll demonstriert.

Wir wollen annehmen, daß der Fremdstoff beim Lösen in ν_r Teilchen der Sorte r, ν_q Teilchen der Sorte q und ν_{23} Teilchen der Sorte (23) dissoziiert, welche die Komponente 2 mit der Komponenten 3 (Inertsalz) gemeinsam hat. Zudem nehmen wir an, daß seine Aktivität a_2 durch das folgende Produkt von Teilchenaktivitäten gegeben ist:

$$a_2 = (a_r)^{\nu_r} \cdot (a_q)^{\nu_q} \cdots (a_{23})^{\nu_{23}} = (m_r)^{\nu_r} \cdot (m_q)^{\nu_q} \cdots (m_{23})^{\nu_{23}} \cdot \gamma_2 \qquad (2.4)$$

117

Der Aktivitätskoeffizient γ_2 des Fremdsalzes wird willkürlich auf die reine gesättigte Inertsalzlösung (Medium) bezogen:

$$\gamma_2 = (\gamma_r)^{\nu_r} \cdot (\gamma_q)^{\nu_q} \cdots (\gamma_{23})^{\nu_{23}} \to 1, \quad \text{wenn } m_2 \to 0 \qquad (2.5)$$

Wenn die Dissoziation vollständig ist, sind die Partikelkonzentrationen durch die folgenden Zusammenhänge gegeben:

$$m_x = m_2 \cdot \nu_x, \, x = r, q, \ldots, \quad \text{und } m_{23} = m_2 \cdot \nu_{23} + m_3 \cdot \nu_{32} \qquad (2.6)$$

ν_{23} ist dabei die Zahl der Ionen, welche die Komponente 3 mit dem Fremdstoff 2 gemeinsam hat.

Nach dem Einsetzen der partiellen Ableitungen, die aus den Ansätzen (2.4) und (2.6) gebildet werden, und den Temperaturkoeffizienten der Aktivitäten (Ausdruck (2.2)), nimmt Gl. (2.1) die folgende Form an:

$$\frac{dT}{dm_2} \cdot \frac{W}{RT^2} = (\nu_r + \nu_q + \cdots)$$
$$+ m_2 \left[\frac{(\nu_{23})^2}{m_{23}} + \beta_{22} + \left(\frac{dm_3}{dm_2} \frac{\nu_{23} \cdot \nu_{32}}{m_{23}} + \beta_{23} \right) \right]$$

$$\text{mit } \beta_{22} = \frac{\delta \ln \gamma_2}{\delta m_2} \text{ und } \beta_{23} = \frac{\delta \ln \gamma_2}{\delta m_3} \qquad (2.7)$$

Diese Gleichung erlaubt in ihrer differentiellen Form nur über den Grenzwert der experimentellen $\Delta T/m_2$-Werte eine Aussage zu machen. Aus dem allgemeinen Verhalten von Elektrolytlösungen (s. (5)) erwartet man, daß beim Übergang $m_2 \to 0$, d. h. $\gamma_2 \to 1$ und damit $\ln \gamma_2 \to 0$, die β-Werte nicht unendlich groß werden. Dann verschwindet nämlich das zweite Glied auf der rechten Seite von Gl. (2.7), und man erhält:

$$\lim_{m_2 \to 0} \left[\frac{\Delta T}{m_2} \right] = \left[\frac{dT}{dm_2} \right]_{m_2 = 0} = \frac{RT^2}{W} \cdot \nu = E \cdot \nu \qquad (2.8)$$
$$\text{mit } \nu = \nu_r + \nu_q + \cdots = \nu_2 - \nu_{23},$$

ν_2 bedeutet die Gesamtzahl der Ionen pro Formel Fremdstoff, während für ν davon nur diejenigen zu zählen sind, welche das Fremdsalz nicht mit dem Inertsalz gemeinsam hat (Fremdionen).

Gl. (2.8) gilt für endliche Konzentrationen m_2 nicht mehr; sie gibt aber leider auch keine Auskunft darüber, nach welchem Gesetz der Meßwert $\Delta T/m_2$ ändert, so daß man auf $m_2 = 0$ extrapolieren könnte. Man kann nun aber über die Neigung der Funktion $\left(\dfrac{\Delta T}{m_2} \right)$ versus m_2 eine Aussage machen, indem man ΔT aus Gl. (2.7) durch Integration ermittelt.

Wenn man annimmt, daß W von T unabhängig ist, sowie m_3 im Meßbereich $m_2 << m_3^0 = m_3$ $(m_2 = 0)$ konstant ist und damit m_{23} den
Betrag $m_{23} \approx m_3^0 \cdot \nu_{32}$ annimmt, sowie daß $dm_3/dm_2 = 0$ ist, ergibt die
Integration zwischen den Werten T_0 (Temperatur des Mediums bei dem
vorgewählten Partialdruck des Wassers, identisch mit Zellentemperatur)
und T bzw. $m_2 = 0$ und m_2 die folgende Gleichung:

$$\frac{\Delta T}{E \cdot m_2} \equiv \nu(m_2) = \nu + m_2 \cdot \frac{(\nu_{23})^2}{2 \cdot \nu_{32} \cdot m_3^0} + \int_0^{m_2} \beta_{22} dm_2 \qquad (2.9)$$

Da das zweite und dritte Glied auf der rechten Seite dieser Gleichung
nicht Null sind, kann nur eine scheinbare Teilchenzahl $\nu(m_2)$ durch Messung von $\Delta T = (T - T_0)$ bestimmt werden.

Die Funktion $\nu(m_2)$, m_2 stellt eine Gerade dar, wenn β_{22} konstant
bzw. gleich Null ist. Der letztgenannte Fall bedeutet, daß der Aktivitätskoeffizient γ_2 von der Konzentration unabhängig ist, d.h., daß sich der
Fremdstoff ideal verhält. Die Neigung N^* der $\nu(m_2)$, m_2-Geraden beträgt
dann:

$$N^* = \nu_{23}^2/(2 \cdot \nu_{32} \cdot m_3^0) \qquad (2.10)$$

Haben Fremdsalz und Inertsalz keine gemeinsamen Ionen, so ist ν_{23}
und damit auch N^* Null. Man erwartet aus dem allgemeinen Verhalten
von Elektrolytlösungen, daß β_{22} dann konstant oder sogar Null wird,
entsprechend einer Geraden für die Funktion $\nu(m_2)$, m_2, wenn der Meßbereich (Variation von m_2) und der Inertelektrolyt so gewählt werden,
daß m_2/m_3^0 möglichst klein bleibt.

Tatsächlich können wir hier experimentell zeigen, daß die scheinbare
Teilchenzahl $\nu(m_2)$, die der Fremdstoff pro Formel liefert, im großen
Bereich von dessen Molalität m_2 linear abhängt. Die Neigung N der expe-
rimentellen $\nu(m_2)$ m_2-Geraden ist aber in den meisten Fällen von N^*
verschieden. Das bedeutet, daß β_{22} nicht Null, aber doch konstant ist.
Gl. (2.9) nimmt daher die folgende einfache Form an:

$$\frac{\Delta T}{E \cdot m_2} \equiv \nu(m_2) = \nu + m_2 \cdot (N^* + \beta_{22}) = \nu + N \cdot m_2 \qquad (2.11)$$

$\beta_{22} \neq 0$, d.h. $N \neq N^*$ bedeutet, daß sich das System nicht ideal verhält. Es ist üblich, die Abweichungen von idealem Verhalten mit Aktivitätskoeffizienten zu beschreiben. Nimmt man nun an, daß die Temperaturabhängigkeit von γ_2 im Meßbereich vernachlässigbar ist, so kann man
die partielle Ableitung β_{22} dem Differentialquotienten $d\ln\gamma_2/dm_2$

gleichsetzen. Dann kann aber der Aktivitätskoeffizient aus folgender Gleichung berechnet werden:

$$(1/m_2) \cdot \log \gamma_2 = (N-N^*)/2{,}303, \text{ wenn } dm_3/dm_2 = 0 \text{ und } \delta ln\gamma_2/\delta T = 0$$

$$(2.12)$$

Schließlich sei noch vermerkt, daß die Gl. (2.9) (bzw. (2.11)) auch die Konzentrationsabhängigkeit der salzkryoskopisch bestimmten scheinbaren Teilchenzahlen wiedergeben müßte. Natürlich muß dann E durch die kryoskopische Konstante ersetzt werden, und ΔT bedeutet dann die Depression des Fixpunktes.

2.2 Die salzebullioskopische Konstante

Um die Vorteile der Salzebullioskopie voll auszunützen, sollte das Inertsalz folgenden Anforderungen genügen:

a) Damit man durch Extrapolation der experimentellen Werte $v(m_2)$ direkt die Nuklearität n erhält $(1/n = v_0)$, soll das Inertsalz das Gegenion G des zu untersuchenden Komplexes (Fremdsalz) enthalten.

b) Das Inertsalz sollte eine mittlere Löslichkeit aufweisen. In sehr konzentrierten Elektrolyten ist wegen Aussalzeffekten die Löslichkeit vieler Komplexsalze nämlich sehr gering. Eine zu geringe Konzentration des Inertsalzes andererseits verkürzt den Gültigkeitsbereich der linearen Gl. (2.11).

c) Das Inertsalz sollte in der festen Phase wasserfrei vorliegen. Die Konzentration des Fremdsalzes berechnet man nämlich zweckmäßig aus der Einwaage. Beim Auflösen oder Auskristallisieren eines Hydrates des Inertsalzes würde aber die Gleichgewichtskonzentration mit der aus der Einwaage berechneten nicht mehr übereinstimmen. Salze, welche aus der Lösung wasserfrei kristallisieren, sind gewöhnlich auch nicht hygroskopisch, was von Vorteil ist.

d) Die Auflösungsgeschwindigkeit des Inertsalzes sollte groß sein, damit die Lösung am Inertsalz immer gesättigt bleibt. Dieses ist besonders wichtig bei Salzen, dessen Löslichkeit stark von der Temperatur abhängt.

Da in der Komplexchemie Komplexkationen besonders oft in Form von Chloriden, Nitraten oder Perchloraten präparativ gefaßt werden und man Komplexanionen als Natrium- oder Kaliumsalze kristallisiert, wäre als Inertelektrolyt ein Alkalichlorid, -nitrat oder -perchlorat erwünscht. Von diesen haben NaCl, KCl und KNO$_3$ mittlere Löslichkeiten und kristallisieren ohne Kristallwasser. Andererseits haben NaNO$_3$ sowie NaClO$_4$ zu große und KClO$_4$ eine zu geringe Löslichkeit. Damit hat man günstige Inertsalze für die Salzebullioskopie beider Alkalisalze komplexer Anionen und auch für Chloride und Nitrate komplexer Kationen. Hingegen fehlt ein salzebullioskopisches Medium für die Komplexsalze mit ClO$_4^-$ als Gegenion. Ammoniumperchlorat hat zwar eine günstige Löslichkeit, aber NH$_4^+$ kann deprotoniert werden, was oft nachteilig ist. Hingegen erfüllt das quaternäre Trimethyl-äthylammoniumperchlorat [(CH$_3$)$_3$C$_2$H$_5$N]ClO$_4$ die Bedingungen a bis d ausgezeichnet. Seine Löslichkeit liegt zwar eher an der unteren Grenze, was sich aber kaum je nachteilig auswirkt.

Die salzebullioskopische Konstante E kann nach der Definitionsgleichung (2.3) berechnet werden, wofür man die Wärme W benötigt, um 1 kg Wasser aus der gesättigten Inertsalzlösung zu verdampfen. Natürlich könnte W prinzipiell direkt kalorimetrisch gemessen werden. Eine einfachere Methode besteht jedoch darin, daß man den Wasserdampfdruck über der gesättigten Lösung in Abhängigkeit der Temperatur bestimmt. Gemäß der Clausius-Clapeyron-Gleichung erhält man dann W aus der Neigung der log p, (1/T)-Geraden:

$$W = 2{,}303 \cdot m_1 \cdot R \cdot (\varDelta(\log p)/\varDelta(1/T)) = 253{,}9 \cdot [\varDelta(\log p/\varDelta(1/T)] \quad (2.13)$$
$$[W] = cal/kg$$

Wenn der Partialdruck p und dessen Änderung mit der Temperatur dp/dT für die gesättigte Inertsalzlösung bei der Arbeitstemperatur bekannt sind, kann E auch folgendermaßen erhalten werden:

$$E = p/m_1 \cdot (dp/dT), \quad m_1 = 55{,}51 \ Mol \ H_2O/kg \ H_2O \quad (2.14)$$

In der folgenden Tabelle sind die Daten für einige Inertelektrolyte zusammengestellt. Für drei der sechs Inertsalzlösungen konnte W aus den Dampfdruckdaten von *Carr* und *Harris* (6) nach Gl. (2.13) berechnet werden. Bei 30 °C (Arbeitstemperatur der verwendeten Meßzelle) beträgt RT^2 = 182,6 kcal · Grad/Mol. Dividiert man diese Zahl mit W, so erhält man E. Zur Berechnung der idealen Neigung N* der $v(m_2)$, m_2-Geraden (s. Gl. (2.10)) benötigt man die Molalität der reinen gesättigten Salzlösung, die in der vierten Zeile angegeben ist. Da für das Trimethyläthyl-ammoniumperchlorat keine Literatur-Werte vorhanden waren, wurden Löslichkeitsbestimmungen ausgeführt. In der letzten Zeile ist noch die prozentuale Temperaturabhängigkeit der Löslichkeit (% pro °C) bei 30 °C angegeben.

Tabelle 2.1. *Salzebullioskopische Konstante und Löslichkeit einiger Inertelectrolyte bei 30 °C*

Inertsalz	KCl	KNO$_3$	NaCl	NaNO$_3$	[Me$_3$EtN]ClO$_4$
W (kcal/kg)	569	—	574	571	—
E (Grad · kg/Mol)	0,321	—	0,318	0,320	—
m_3^0 (Mol/kg)	5,00	4,54	6,18	11,36	0,890
($\Delta m_3^0/\Delta$T) (100/m_3^0)	0,80	3,58	0,79	0,84	4,48

Die Verdampfungswärme W für 1 kg Wasser aus den gesättigten Salzlösungen weicht nicht stark ab von der Verdampfungswärme des reinen Wassers (581 kcal/kg bei 30 °C). Die salzebullioskopischen Konstanten sind daher etwa gleich groß wie die ebullioskopische Konstante des Wassers.

2.3 Erniedrigung des Dampfdruckes über einer gesättigten Inertsalzlösung durch ein Fremdsalz

Zwei gesättigte Inertsalzlösungen mit festem Inertsalz als Bodenkörper, von denen die eine noch etwas Fremdstoff enthält, besitzen bei gleicher Temperatur verschiedene Dampfdrucke. Die Dampfdruckdifferenz Δp soll nun in Abhängigkeit der Konzentration des Fremdstoffes untersucht werden.

Die Bedingung für das thermodynamische Gleichgewicht lautet in diesem 3-Phasen-System wie folgt:

$$\frac{dp}{dm_2}\left(m_1\frac{\delta \ln a_1'}{\delta p} - m_1\frac{\delta \ln a_1}{\delta p} + m_3\frac{\delta \ln a_3''}{\delta p} - m_3\frac{\delta \ln a_3}{\delta p}\right) =$$

$$= m_2\frac{\delta \ln a_2}{\delta m_2} + m_2\frac{dm_3}{dm_2}\frac{\delta \ln a_2}{\delta m_3} \qquad (2.15)$$

p bedeutet den Partialdruck des Wassers. Die übrigen Symbole haben die gleiche Bedeutung wie in Gl. (2.1).

Die linke Seite dieser Gleichung wird mit Hilfe der folgenden Beziehung umgeformt:

$$\left[\frac{\delta \ln a_i}{\delta p}\right]_T = \frac{\bar{V}_i - \bar{V}_i^0}{RT}, \qquad (2.16)$$

wobei $\bar{V}_i$ das partiale Molvolumen der Komponente i bedeutet. An der rechten Seite der Gl. (2.15) wird gleichzeitig die Aktivität des Fremdsalzes durch die unabhängige Variable m_2 gemäß Ansatz (2.4) ersetzt. So erhält man den folgenden Wert für die Dampfdruckdifferenz des Mediums mit und ohne Fremdsalz:

$$\frac{dp}{dm_2}\left(m_1 \frac{\bar{V}_1' - \bar{V}_1}{RT} + m_3 \frac{\bar{V}_3'' - \bar{V}_3}{RT}\right) = \nu + m_2 \left[\beta_{22} + \right.$$

$$\left. + \frac{(\nu_{23})^2}{m_{23}} + \frac{dm_3}{dm_2}\left(\frac{\nu_{23} \cdot \nu_{32}}{m_{23}} + \beta_{23}\right)\right] \qquad (2.17)$$

Die Änderung des partialen Molvolumens vom Inertsalz bei Überführen aus der gesättigten Lösung in die reine feste Phase $(\bar{V}_3'' - \bar{V}_3)$ ist vernachlässigbar klein im Vergleich zu der Änderung des partialen Molvolumens vom Wasser $(\bar{V}_1' - \bar{V}_1)$. In erster Näherung kann man sogar das partiale Molvolumen des Wassers in der Lösung $\bar{V}_1$ gegenüber dem Molvolumen des Dampfes $\bar{V}_1'$ vernachlässigen $(\bar{V}_1 << \bar{V}_1')$. Aus Gl. (2.15) kann nun $\Delta p = p_0 - p$ durch Integration erhalten werden. Dabei wird wiederum angenommen, daß dm_3/dm_2 vernachlässigbar klein und β_{22} konstant ist.

$$\frac{(p_0 - p) \cdot m_1}{p_0 \cdot m_2} = \nu_p(m_2) = \nu + \left(\frac{(\nu_{23})^2}{2 \cdot m_{23}} + \beta_{22}\right) \cdot m_2 \qquad (2.18)$$

Gemäß Gl. (2.18) hängt die scheinbare Fremdteilchenzahl $\nu_p(m_2)$ von der Konzentration des Fremdstoffes linear ab. Die Ausdehnung des linearen Bereiches, d.h. die Grenzen der Anwendbarkeit von Gl. (2.18) können aber nur experimentell ermittelt werden. Man kann nämlich im voraus nicht sagen, bis zu welcher Konzentration die oben gemachten Annahmen zutreffen.

2.4. Zusammenhang zwischen der isothermen Dampfdruckerniedrigung und der isobaren Temperaturerhöhung

Die Aussagen der Gl. (2.11) und (2.18) können folgendermaßen zusammengefaßt werden: wenn man einer gesättigten Inertsalzlösung mit festem Inertsalz als Bodenkörper (Medium) ein Fremdsalz in einer Konzentration von m_2 zusetzt, so verändert sich die Aktivität der flüchtigen Komponente 1 (Wasser). Diese Änderung kann entweder als Dampfdruckdifferenz, zwischen den Medien mit und ohne Fremdsalz, die beide dieselbe Temperatur T besitzen (isotherme Dampfdruckerniedrigung Δp), oder als Temperaturdifferenz, zwischen den Medien mit und ohne Fremdsalz, wenn sie unter gleichem Dampfdruck p stehen (isobare Temperaturerhöhung, ΔT), beobachtet werden. Man kann entweder den einen oder den andern dieser Effekte benutzen um eine Information über die Anzahl $\nu(m_2)$ der Fremdpartikeln zu erhalten, die pro Formel des Fremdstoffes in das Salzmedium hineingelangen:

$$\nu(m_2) = \frac{(p_0-p)m_1}{p_0 \cdot m_2} = \frac{(T-T_0)}{E \cdot m_2} \tag{2.19}$$

3. Experimentelle Bestimmung der scheinbaren Teilchenzahl $\nu(m_2)$

3.1. Die Hill'sche Methode (3, 7—10)

Die ursprüngliche Methode besteht darin, daß man kleine Proben der Lösung und des Lösungsmittels in eine mit dem Lösungsmittel gesättigte Atmosphäre bringt. Es stellt sich dann eine über längere Zeit konstant bleibende Temperaturdifferenz zwischen beiden Proben ein. Diese Temperaturdifferenz entsteht dadurch, daß der Partialdruck an der Oberfläche der Lösung kleiner ist als in der Atmosphäre, was eine Kondensation von Lösungsmittelmolekeln nach sich zieht, wobei wegen der frei werdenden Kondensationsenthalpie eine Erwärmung dieser Probe stattfindet. Das System erreicht bald einen quasi-stationären Zustand, bei dem die durch Kondensation der Lösung zugeführte Wärme durch Wärmeverluste infolge Wärmeleitung und Abstrahlung kompensiert wird. So ist es möglich, sehr kleine Partialdruckunterschiede durch Messung stationärer Temperaturdifferenzen recht genau zu ermitteln. Die Hill-Methode findet in letzter Zeit bei der Bestimmung von Molekulargewichten organischer Verbindungen verbreitet Anwendung (siehe (7)). Wie gezeigt werden soll, ist sie auch mit gesättigten Salzlösungen als Lösungsmittel anwendbar.

Bringt man zwei Proben von gesättigten wäßrigen Lösungen eines Inertelektrolyten mit festem Salz als Bodenkörper (Medium), deren eine neben dem Inertsalz noch etwas gelöstes Fremdsalz enthält, in die gemeinsame Atmosphäre, deren Wasserdampfdruck demjenigen des Mediums entspricht, so ist zu erwarten, daß sich zwischen den beiden Proben ein Temperaturunterschied ausbilden wird. Im thermodynamischen Gleichgewicht können die beiden Lösungen nach Gl. (2.11) nur dann sein, wenn diejenige mit dem Fremdstoff um ΔT wärmer ist als diejenige ohne Fremdstoff, da über beiden Proben derselbe Wasserdampfdruck lastet. Diese Temperaturerhöhung kommt wieder durch Kondensation von Wasser an der Oberfläche des Mediums mit dem Fremdstoff zustande. Natürlich ist nicht zu erwarten, daß der sich einstellende Temperaturunterschied einem thermodynamischen Gleichgewicht entspricht und die Gl. (2.11) anwendbar ist. Die Versuche werden ja nicht unter adiabatischen Verhältnissen ausgeführt. Die Probe höherer Temperatur wird durch Abstrahlung und Konvektion Wärme verlieren. Aber dieser Wärmeverlust ist für kleine Temperaturunterschiede selber wieder ΔT proportional. Das Verhältnis des tatsächlichen, quasi-stationären Temperaturunterschiedes zum theoretischen Unterschied nach Gl. (2.11) wird daher eine Konstante ergeben, welche als *Temperaturausbeute* η bezeichnet werden soll. Die salzebullioskopische Grundgleichung für die Hill-Methode lautet dann:

$$\Delta T_s = \eta \cdot \Delta T_t = \eta \cdot E \cdot m_2 \cdot \nu(m_2) \qquad (3.1)$$

Da die Temperaturausbeute η nicht nur von den thermodynamischen Eigenschaften des Lösungsmittels, sondern auch von der Konstruktion der Meßzelle abhängt, kann die Konstante $\eta \cdot E$, welche in Gl. (1) und (3.2) mit K bezeichnet ist, nur mit Hilfe einer Eichsubstanz bestimmt werden. Gemäß Gl. (2.11) hängt $\nu(m_2)$ von m_2 linear ab, so daß K aus dem Ordinatenabschnitt der $(\Delta T_s/m_2)$, m_2-Gerade bestimmt werden kann.

$$\Delta T_s/m_2 = K \cdot \nu_0 + K \cdot N \cdot m_2 \qquad (3.2)$$

Als Eichsubstanzen dienten:

KCl für die Messungen mit Inertsalz NaCl, NaNO$_3$, $(\nu_0 = 1)$
KNO$_3$ für die Messungen mit Inertsalz NaNO$_3$, $(\nu_0 = 1)$
NaCl für die Messungen mit Inertsalz Me$_3$EtN ClO$_4$, $(\nu_0 = 2)$

Die Messungen wurden nach einer früher beschriebenen Technik (*4*) ausgeführt, wobei aber eine verbesserte Apparatur (*22*) verwendet wurde.

3.2. Die isopiestische Methode der Dampfdruckmessung

Die isopiestische oder isotonische Methode für die Messung des Wasserpartialdruckes über Lösungen hat *Bousfield* (*19*) im Jahre 1918 eingeführt. Sie besteht in der Ermittlung der Konzentration einer Lösung, welche man über die Dampfphase mit einer Eichlösung ins thermodynamische Gleichgewicht gebracht hat. Der Dampfdruck der Eichlösung kann dann aus Dampfdrucktabellen erhalten werden.

Die Gleichgewichtskonzentration m beider Lösungen kann entweder analytisch bestimmt, oder aus der Anfangskonzentration m^0 und der Gewichtsänderung Δg, welche die Lösung während des Equilibrierens erleidet, berechnet werden.

Als Eichlösung verwendet man gewöhnlich Lösungen von NaCl, KCl, CaCl$_2$ und H$_2$SO$_4$, deren Dampfdruck genau bekannt und in Abhängigkeit der Molalität m tabelliert ist (20). Dabei werden relative Dampfdrucke, d. h. die Aktivitäten a_w des Wassers angegeben, die man auf den Dampfdruck des reinen Wassers (23,753 mmHg bei 25 °C) bezogen hat:

$$a_w = p/23{,}753 \qquad (3.3)$$

Die Ausführung isopiestischer Messungen ist denkbar einfach. Man thermostatiert eine Lösung der Eichsubstanz und die Lösungen, deren Dampfdruck ermittelt werden soll, in einer geschlossenen Meßzelle so lange, bis keine Gewichtsänderungen mehr festzustellen sind. Die Gefäße für die Lösungen sollen gute Wärmeleiter sein. Gewöhnlich stellt man Metallschalen auf einen Kupferblock, damit möglichst keine Temperaturgefälle zwischen den Lösungen bestehen. Die Meßzelle wird dann von Luft befreit und in einen Thermostaten versenkt, dessen Temperatur bei dem gewünschten Wert konstant gehalten wird. Um die Einstellung des Gleichgewichtes zu fördern, wird die Zelle bewegt.

Für die Messungen dieser Arbeit verwendeten wir eine Methode, die ähnlich wie die von *Scatchard*, *Hamer* und *Wood* (21) beschriebene ist.

4. Resultate und Diskussion

4.1. Gültigkeitsbereich

Die Gültigkeit des salzebullioskopischen Grenzgesetzes (2.11) wurde mit gegen 40 verschiedenen Inertsalz-Fremdsalz-Systemen experimentell geprüft. Die Resultate für Systeme mit Inertsalzen NaCl, NaNO$_3$ und KNO$_3$ wurden bereits veröffentlicht (4) und haben die Gültigkeit des Grenzgesetzes (2.11) bis zu hohen Konzentrationen (ca. 1 Mol/kg) bestätigt.

Abb. 4.1 zeigt einige Resultate neuer Messungen mit dem Inertsalz Trimethyläthyl-ammoniumperchlorat. Die scheinbare Teilchenzahl $\nu(m_2)$

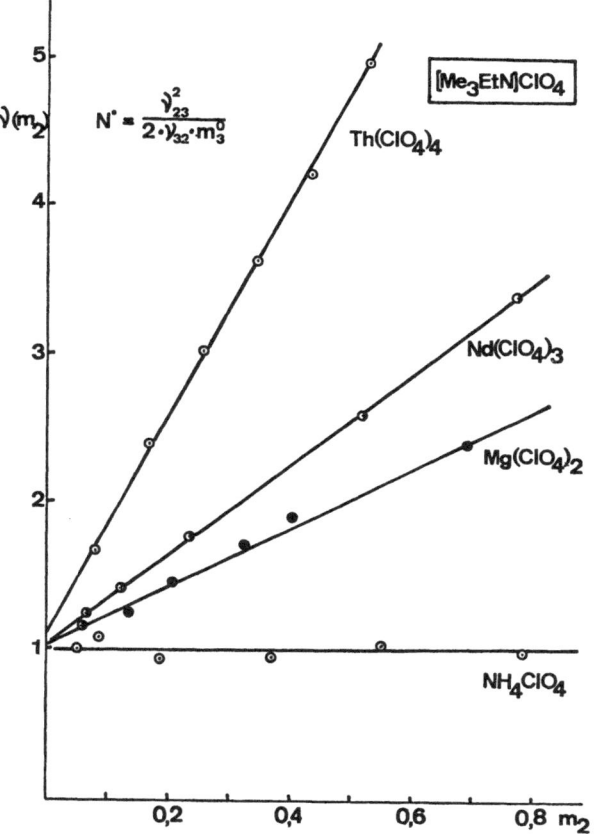

Abb. 4.1. Einfluß der Ladung ($=\nu_{23}$) auf die Neigung N der $\nu(m_2)$, m_2-Geraden

zeigt in diesem Medium ebenfalls eine lineare Abhängigkeit von der Molalität m_2 des Fremdstoffes. Die durch lineare Regression ermittelten Werte für ν_0 und für die Neigung N der $\nu(m_2)$, m_2-Geraden sind zusammen mit den zugehörigen Standardabweichungen $s(\nu_0)$ und $s(N)$ in der Tabelle 4.1 zusammengestellt.

Um die Gültigkeit von Gl. (2.11) zu prüfen, müssen die Abweichungen ($\nu_0-\nu$) näher betrachtet werden. Wir wollen festlegen, daß die Gl. (2.11) nur dann gültig ist, wenn der absolute Betrag von ($\nu_0-\nu$) die Fehlerbreite von ν_0 sicher nicht überschreitet. Formelmäßig lautet diese Bedingung:

$$\left|(\nu_0-\nu)\right| < t \cdot s(\nu_0) ,$$

B. Magyar

wobei t den Wert der t-Funktion *(11)* für 99%ige Sicherheit und für einen Freiheitsgrad von $(z-2)$ bedeutet. Die Anzahl der Meßpunkte z ist in der Tabelle 4.1 ebenfalls angegeben.

Alle ν_0-Werte stimmen mit der wahren Fremdteilchenzahl innerhalb der Fehlergrenzen überein, so daß die Gültigkeit der Gl. (2.11) in angegebenem Konzentrationsbereich auch für dieses System bestätigt ist.

Tabelle 4.1. *Zusammenstellung der Resultate für das System mit* $[Me_3EtN]ClO_4$ *als Inertsalz*

Fremdsalz	ν	ν_0	N	z	$s(\nu_0)$	$s(N)$	Konzentr. Bereich für m_2	$(1/m_2)$ $\cdot \log \gamma_2$
NH_4ClO_4	1	0,983	+0,061	7	0,044	0,080	0,050—0,981	—0,174
$Mg(ClO_4)_2$	1	1,038	+1,980	6	0,109	0,305	0,066—0,692	—0,116
$Nd(ClO_4)_3$	1	1,027	+3,057	8	0,046	0,164	0,042—0,776	—0,869
$Th(ClO_4)_4$	1	1,110	+7,255	6	0,050	0,144	0,085—0,531	—0,753
Me_3EtNOH	1	0,997	+0,571	5	0,032	0,046	0,197—1,088	+0,047
$HClO_4$	1	1,037	+0,244	4	0,047	0,086	0,154—0,805	—0,094
$Me_3EtNH_2PO_4$	1	0,932	+0,477	6	0,015	0,028	0,127—0,840	+0,007
$NaNO_3$	2	1,931	—0,469	6	0,070	0,179	0,064—0,840	—0,202

Sehr interessant ist, daß die Neigung der experimentellen Geraden für verschiedene Inertsalze mit Hilfe der einfachen Beziehung (2.10) vorausgesagt werden kann. Nach dieser Beziehung sollten die Neigungen in Systemen mit dem gleichen Inertsalz und verschiedenen Fremdsalzen mit der Anzahl ν_{23} der gemeinsamen Ionen pro Formel Fremdstoff quadratisch zunehmen. Daß dem tatsächlich so ist, ist aus Abb. 4.2 ersichtlich, wo die experimentelle Neigung für die Perchlorate und Nitrate von Na^+, Mg^{2+}, Nd^{3+} und Th^{4+} gegen ν_{23}^2 aufgetragen worden sind. Als Inertsalz diente $Me_3EtNClO_4$ $(m_3^0 = 0,890)$ für die Perchlorate und $NaNO_3$ $(m_3^0 = 11,36)$ für die Nitrate. Für konstant bleibendes ν_{23} sollten in Systemen mit verschiedenen Inertsalzen, nach Gl. (2.10), die N-Werte mit $1/m_3^0$ linear ansteigen. Wie Abb. 4.3 zeigt, trifft auch diese Voraussage zu. Es sind also große N-Werte zu erwarten, wenn das Fremddion hochgeladen ist (dann ist die Zahl der Gegenionen ν_{23}, die das Fremdsalz mit dem Inertsalz gemeinsam hat, groß) und wenn die Konzentration des Inertsalzes klein ist (geringe Löslichkeit des Inertsalzes).

Es muß jedoch nochmals betont werden, daß diese Voraussagen nur für ein ideales oder beinahe ideales System zutreffen können. Man beachte, daß aus der Normierung der Aktivitätskoeffizienten (s. Gl. (2.5))

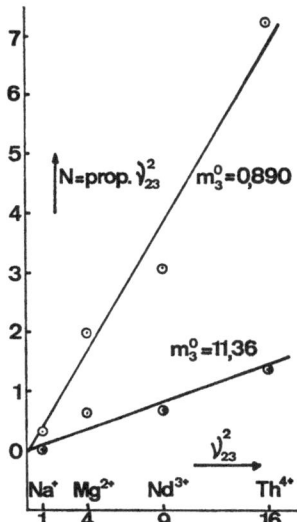

Abb. 4.2. Abhängigkeit der N-Werte von ν_{23}

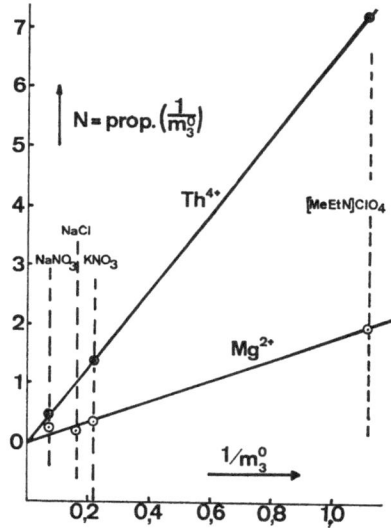

Abb. 4.3. Abhängigkeit der N-Werte von m_3^0

lediglich folgt, daß $\gamma_2 = 1$ wird bei $m_2 = 0$, nicht aber, daß auch β_{22} verschwindet. Es ist deshalb keine thermodynamische Notwendigkeit, daß die Neigung einer experimentellen $\nu(m_2), m_2$-Gerade bei $m_2 = 0$ mit N* übereinstimmt (Gl. (2.9)). Es scheint ja auch sehr unwahrscheinlich, daß die Geraden mit negativer Neigung bei sehr kleinen, für die Messung nicht mehr zugänglichen Konzentrationen, in eine Kurve und anschließend nochmals in eine Gerade mit der positiven Neigung N* übergehen würden. Wie erwähnt, bedeutet $N \neq N*$ nichtideales Verhalten. Es ist üblich, die Abweichungen vom idealen Verhalten mit der Angabe von Aktivitätskoeffizienten zu beschreiben. Darum wurden Aktivitätskoeffizienten aus $(N-N*)$ mit Hilfe von Gl. (2.12) in Form des Ausdruckes $(1/m_2) \log \gamma_2$ berechnet und in der letzten Kolonne der Tabelle 4.1 angegeben. Sie sind natürlich nur für den untersuchten Konzentrationsbereich gültig.

4.2. Einheitliche Darstellung von salzebullioskopischen und salzkryoskopischen Resultaten

Wie erwähnt, müßte Gl. (2.11) auch die Konzentrationsabhängigkeit der salzkryoskopisch bestimmten Teilchenzahlen wiedergeben. Natürlich muß E dann durch die kryoskopische Konstante K ersetzt werden, und ΔT bedeutet die Depression des Fixpunktes.

Als Beispiel seien die Messungen von *Jahr* und *Wegener* (14) in dieser Form dargestellt (s. Abb. 4.4). Diese Autoren untersuchten das kryoskopische Verhalten von NaCl und $Na_4[Fe(CN)_6]$ in schmelzendem

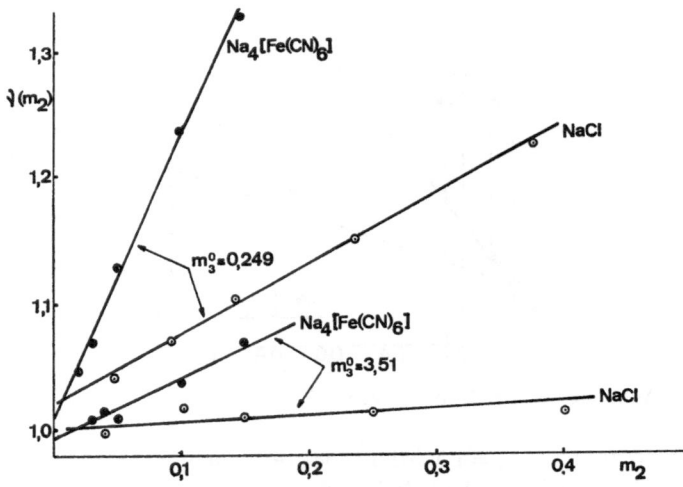

Abb. 4.4. Salzkryoskopisch bestimmte, scheinbare Teilchenzahlen

„Glaubersalz" ($m_3^0 = 3{,}51$) und im $Na_2SO_4 \cdot 10\ H_2O$-Eis-Eutektikum ($m_3^0 = 0{,}249$). Die größere Neigung der $\nu(m_2), m_2$-Geraden für das Eutektikum ist gemäß Gl. (2.10) auf die niedrigere Konzentration des Inertsalzes zurückzuführen. Diese Gleichung sagt sowohl im Eutektikum als auch in der Glaubersalzschmelze steilere Geraden für das hochgeladene $Fe(II)(CN)_6$-Fremdion voraus, was ebenfalls experimentelle Bestätigung findet.

Eine einheitliche Wiedergabe von salzkryoskopischen und salzebullioskopischen Meßresultaten ist also mit Hilfe der Gl. (2.11) möglich, und wegen besserer Vergleichbarkeit wünschenswert.

4.3. Salzebullioskopische Messungen mit DFA als Komponente 1

Bei der Ableitung des Grenzgesetzes (2.11) wurde von der Komponente 1, außer ihrer Flüchtigkeit, keine besondere Eigenschaft verlangt. Danach könnte man prinzipiell ein beliebiges molekulargebautes Lösungsmittel einsetzen. Sind nun die Ansätze (2.4), (2.5) und (2.6) gültig, so sollte das Grenzgesetz (2.11) auch für nichtwäßrige Systeme gelten. Um dieses zu prüfen, wurde in einer Versuchsserie Dimethylformamid (DFA) als Komponente 1 eingesetzt. Als Inertsalz diente $Me_3EtNClO_4$. In diesem System sind nun DFA, Me_3EtN^+ und ClO_4^- keine Fremdpartikeln. Gemäß Gl. (2.11) sollten daher das $Nd(ClO_4)_3$ und das molekulargebaute Triphenylphosphin die gleiche Fremdteilchenzahl $\nu_0 = 1$ ergeben. Das ist auch gefunden worden, wie die Abb. 4.5 zeigt. Der Verlauf der $\nu(m_2)\,m_2$-

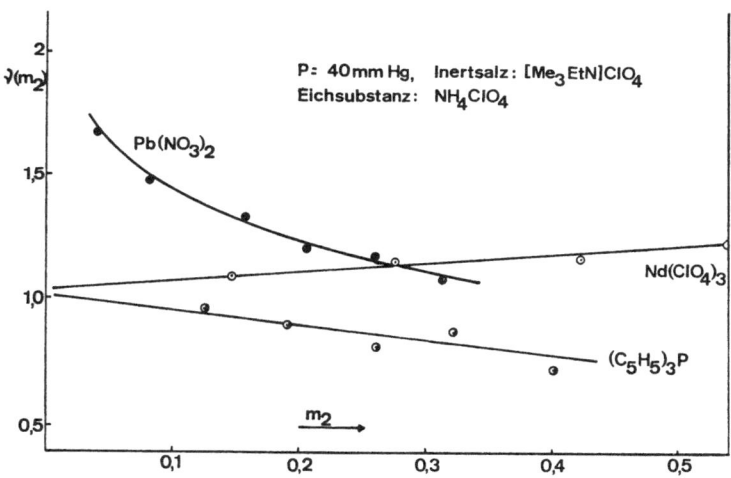

Abb. 4.5. Salzebullioskopische Messungen in DFA

B. Magyar

Kurve für $Pb(NO_3)_2$ kann jedoch mit Hilfe des Grenzgesetzes (2.11) nicht interpretiert werden. Hier handelt es sich vermutlich um Ionenassoziation, so daß der Ansatz (2.6) nicht mehr erfüllt ist. Man könnte solche Kurven mit Ionengleichgewichten beschreiben, worauf hier jedoch verzichtet wird.

Die Salzebullioskopie und die Salzkryoskopie sind nämlich zum Studium von Ionengleichgewichten wenig geeignet, da der zur Verfügung stehende Konzentrationsbereich sehr eng ist.

4.4. Anwendung der Salzebullioskopie zur Bestimmung der Nuklearität von Komplexionen

Bei Komplexsalzen kennt man im Prinzip die Ionengewichte ihrer Bausteine erst nach der Ermittlung der Nuclearität. Mit m_2 bezeichnet man daher die Molalität des Zentralatoms M. Diese kann natürlich aus der Menge der eingewogenen Substanz und derjenigen des für die Lösung verwendeten Wassers berechnet werden, insofern die stöchiometrische Zusammensetzung des zu untersuchenden Salzes bekannt ist. Für die Untersuchung der Nuclearität verwendet man ein Inertsalz, bei dem eines seiner Ionen mit dem Gegenion des Komplexions übereinstimmt. Die erhaltenen ν_0-Werte pro M lassen sich dann folgendermaßen interpretieren:

$\nu_0 > 1$ bedeutet, daß das Komplexion beim Lösen dissoziiert;
$\nu_0 = 1$, so ist der Komplex mononuclear;
$\nu_0 < 1$ zeigt einen polynuclearen Komplex an.

Die Nuclearität beträgt:

$$n = 1/\nu_0 \pm (1/\nu_0^2) \cdot s(\nu_0) \tag{4.1}$$

In der folgenden Tabelle 4.2 sind Beispiele angegeben, wo die Methode zur Bestimmung der Nuclearität erfolgreich angewandt werden konnte.

Tabelle 4.2. *Salzebullioskopisch bestimmte Nuclearität n einiger Komplexsalze*[a])

No.	Substanzformel	m_2	Inertsalz	n
1	$[Co(dien)(OH)_{3/2}](ClO_4)_{3/2}$	0,08—0,75	$Me_3EtNClO_4$	$2,16 \pm 0,07$
2	$[Cd(tgl)_{8/5}](ClO_4)_{2/5}$	0,13—0,86	$Me_3EtNClO_4$	$10,7 \pm 1,1$
3	$[Ti(dipic)O_{5/2}]Na$	0,03—0,09	NaCl	$1,7 \pm 0,1$
4	$[CrOH(NH_3)_{9/2}]Cl_2$	0,05—0,73	NaCl	$2,04 \pm 0,08$
5	$[SbC_4H_4O_7]Na$	0,18—0,98	$NaNO_3$	$1,89 \pm 0,02$

[a]) Der Diäthylentriamin-Co-Komplex wurde von *J. Bösch*, der Thioglykolato-Cd-Komplex von *K. Gautschi* und der Dipicolato-Ti-Komplex von *J. Mühlebach* hergestellt. Der Autor möchte an dieser Stelle für die erhaltenen Proben danken.

Der Komplex No. 1 ist sicher dinuclear. Der Cd-Komplex No. 2, welcher in der festen Phase aus dekanuclearen Einheiten besteht (15), enthält auch in der Lösung höchst wahrscheinlich dekanucleare Komplexionen. Die Meßresultate sind auch in Abb. 4.6 graphisch dargestellt. Die Messungen an $HClO_4$ und Me_3EtNOH sind in der gleichen Figur wiedergegeben und zeigen, daß H^+ und OH^- ebenfalls als Fremdionen anzusehen sind. Die Messungen an den beiden Komplexsalzen liefern also auch die zusätzliche Information, daß die beiden Komplexionen mit Wasser in keine merkliche Säure-Base-Reaktion eingehen. Der Peroxotitan-Komplex No. 3 ist in der Lösung nur begrenzt haltbar und zudem schlecht löslich.

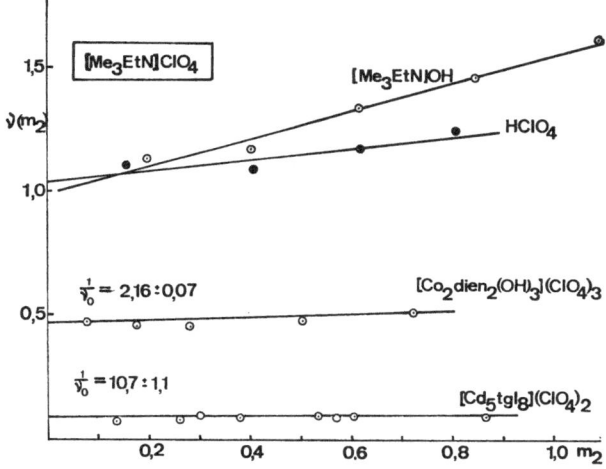

Abb. 4.6. Bestimmung der Nuclearität von Cd- bzw. Co-Komplex

Das wirkte sich auf die Messungen sehr nachteilig aus. Die Existenz eines dinuclearen Komplexions kann trotzdem als bewiesen betrachtet werden. Salz No. 4 ist ein Chrom-Komplex (Hydroxo-Erythro-Chrom-(III)-Chlorid), der sicher dinuclear ist (16), was mit dem gefundenen n-Wert wiederum bestätigt wird. Die Messungen an der Lösung vom Brechweinstein in gesättigter $NaNO_3$-Lösung ergaben einen ν_0-Wert von 1,53. Nun wurde das Kalium-Ion mit Hilfe eines Kationentauschers durch das Natrium-Ion ersetzt. Die Messungen an dem eingeengten Eluat ergaben $\nu_0 = 0,529$, da das Na^+ nicht aktiv ist. Damit ist eindeutig bewiesen, daß der Sb-Komplex im Brechweinstein die Nuclearität 2 besitzt ($n = 1,89$). Zu der gleichen Folgerung gelangten auch *Anderegg* und *Malik* (17) aufgrund neuerer potentiometrischer Messungen.

Die letztgenannte Untersuchung stellt ein Beispiel für die kombinierte Anwendung von Salzebullioskopie und einer Ionentauscher-Methode dar. In diesem Fall bereitete die Umwandlung des K-Salzes in eine Na-Salzlösung keine Schwierigkeiten. Sollte sich diese aber nicht leicht durchführen lassen, so wäre es vorteilhafter, das Inertsalz zu wechseln. Die gleiche Information ($n = 2$) würde man natürlich aus salzebullioskopischen Messungen am Brechweinstein mit KNO_3 als Inertsalz direkt erhalten.

4.5. Resultate und Interpretation der isopiestischen Dampfdruckmessungen

Um die Gültigkeit der Gl. (2.18) zu prüfen, wurden isopiestische Messungen an der Lösung eines Inertsalzes ausgeführt, dessen kleine Löslichkeit eine große Neigung N_p der $v_p(m_2)$, m_2-Geraden erwarten ließ. Dann kann nämlich die Neigung genauer bestimmt werden, was einen besseren Vergleich mit den Voraussagen der Theorie ermöglicht. Da das Trimethyl-ethyl-ammoniumperchlorat diese Voraussetzungen erfüllt und salzebullioskopisch untersucht wurde, wählte man es wiederum als Inertsalz. Die Resultate sind in den Tabellen 4.3 und 4.4 zusammengestellt und in den Abb. 4.7 und 4.8 illustriert.

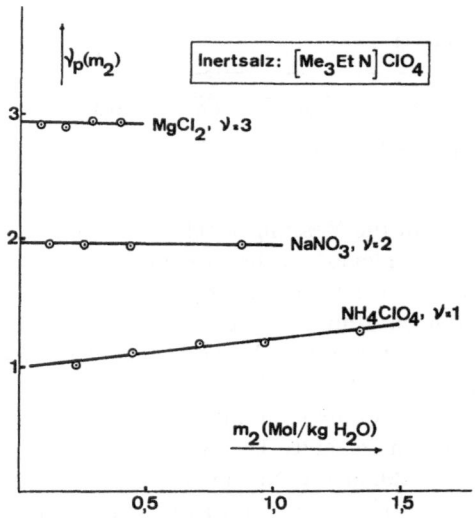

Abb. 4.7. Abhängigkeit der aus Dampfdruckerniedrigung bestimmten scheinbaren Teilchenzahl der Fremddionen $v_p(m_2)$ von der Konzentration des Fremdstoffes m_2

Tabelle 4.3. *Dampfdruck von gesättigten* [Me₃EtN]ClO₄-Lösungen, welche noch einen *Fremdstoff in der Konzentration* m_2 *enthalten (*m_{NaCl} *ist die Konzentration der NaCl-Lösung, welche jeweils mit der binären Lösung isopiestisch ist)*

m_{NaCl}	$p(mmHg)$	m_2			$v_p(m_2)$		
		NH_4ClO_4	$NaNO_3$	$MgCl_2$	NH_4ClO_4	$NaNO_3$	$MgCl_2$
0,5801	23,2995	0	0	0	—	—	—
0,7003	23,2045	0,223	0,116	0,078	1,01	1,94	2,90
0,8433	23,0912	0,446	0,254	0,172	1,11	1,95	2,88
1,0188	22,9509	0,709	0,431	0,283	1,17	1,93	2,94
1,1830	22,8190	0,969	—	0,393	1,18	—	2,91
1,4701	22,5819	1,341	0,877	—	1,27	1,95	—

Mit den Fremdsalzen NH_4ClO_4, $NaNO_3$ und $MgCl_2$ werden 1, 2 und 3 Fremdionen pro Formel Fremdstoff in die gesättigte Lösung von [Me₃EtN]ClO₄ eingeführt. Aus der Dampfdruckerniedrigung der Inertsalzlösung erhielt man v_p-Werte, welche mit der wahren Fremdteilchenzahl v innerhalb der Fehlergrenzen übereinstimmen (s. Abb. 4.7).

Tabelle 4.4 *Dampfdruck von gesättigten* [Me₃EtN]ClO₄-Lösungen, welche je ein *Fremdion in der Konzentration* m_2 *enthalten (*m_{NaCl} *ist die Konzentration der NaCl-Eichlösung, welche jeweils mit den binären Lösungen isopiestisch ist)*

m_{NaCl}	p	m_2			$v_p(m_2)$		
		$NaClO_4$	$Mg(ClO_4)_2$	$Th(ClO_4)_4$	Na^+	Mg^{2+}	Th^{4+}
0,5801	23,2995	0	0	0	—	—	—
0,6327	23,2580	0,0992	—	0,0530	0,997	—	1,855
0,6374	23,2544	0,1018	—	0,0626	1,055	—	1,718
0,6453	23,2480	—	0,1046	0,0714	—	1,189	1,740
0,7266	23,1839	0,2542	—	0,1318	1,084	—	2,089
0,7317	23,1798	0,2494	0,2029	0,1307	1,143	1,405	2,183
0,7903	23,1333	0,3620	—	0,170	1,106	—	2,359
0,8989	23,0468	0,5204	0,3378	0,2178	1,153	1,776	2,754
1,0087	22,9589	0,6258	0,4175	0,2703	1,297	1,945	3,003
1,1979	22,8155	0,8229	0,5317	0,3338	1,353	2,169	3,456
1,3873	22,6509	1,2012	0,6380	0,4008	1,286	2,422	3,855
1,3887	22,6508	1,2719	0,6371	0,3907	1,215	2,425	3,957
1,4576	22,5938	1,2460	0,6602	0,4087	1,349	2,546	4,114
1,5344	22,5273	1,3873	0,7066	0,4415	1,327	2,604	4,167
1,6117	22,4632	1,4928	0,7450	0,4463	1,334	2,674	4,464

Die Neigung der $\nu_p(m_2)$, m_2-Geraden sollte nach Gl. (2.18) Null betragen, wenn das Fremdsalz und das Inertsalz kein gemeinsames Ion enthalten ($\nu_{23} = 0$) und wenn sich der Fremdstoff ideal verhält ($\beta_{22} = 0$). Die Messungen an $NaNO_3$ und $MgCl_2$ bestätigen dieses.

Vergleicht man schließlich die Abb. 4.1 und 4.8, so wird die ähnliche Abhängigkeit der salzebullioskopisch bzw. isopiestisch bestimmten scheinbaren Fremdteilchenzahl von der Konzentration offensichtlich.

Die zahlenmäßige Übereinstimmung der Kenngrößen ν^0 bzw. N, welche nach der salzebullioskopischen und isopiestischen Methode bestimmt wurden, ist aus der folgenden Tabelle 4.5 ersichtlich.

Tabelle 4.5. *Vergleich der salzebullioskopischen und isopiestischen (Index p) Resultate für 3 Systeme mit* $[Me_3EtN]ClO_4$ *als Inertsalz*

Fremdstoff	NH_4ClO_4	$Mg(ClO_4)_2$	$Th(ClO_4)_4$
Wahre Fremdteilchenzahl:	1	1	1
ν^0:	0,98	1,04	1,11
ν_p^0:	0,98	0,97	ca. 1,0
N:	+0,06	+1,98	+7,26
Np:	+0,24	+2,36	+7,43

Die leichte Krümmung der $\nu_p(m_2)$, m_2-Funktion (s. Abb. 4.8) für das Thoriumperchlorat deutet auf eine Hydrolyse hin, was durch die stark saure Reaktion des Salzes in Wasser glaubhaft ist. Um ν_p^0 und N_p zu ermitteln, extrapolierte man auf die Konzentration Null mit Hilfe des gradlinigen Verlaufes bei großen Konzentrationen, wo die Hydrolyse weniger ins Gewicht fällt.

Die Übereinstimmung der N-Werte ist gut. Nach Gl. (2.10) sollte man für N_p etwas größere Werte erhalten, da diese bei tieferer Temperatur (25 °C) und dementsprechend bei kleinerer Konzentration des Inertsalzes bestimmt wurden. N_p/N sollte theoretisch m_3^0 (30 °C)$/m_3^0$ (25 °C) betragen, wobei m_3^0 die Löslichkeit des Inertsalzes bei der betreffenden Temperatur bedeutet. Das Verhältnis der Löslichkeiten beträgt 1,13, was mit dem Verhältnis der N-Werte, die mit Mg^{2+} und Th^{4+} erhalten worden sind, recht gut übereinstimmt.

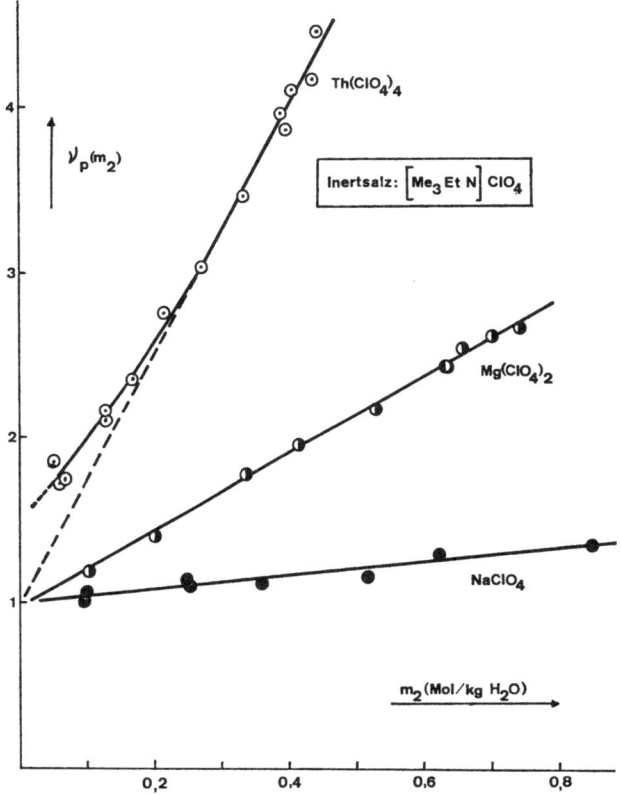

Abb. 4.8. Abhängigkeit der aus Dampfdruckerniedrigung ermittelten scheinbaren Anzahl Fremdionen $\nu_p(m_2)$ von der Konzentration m_2

5. Zusammenfassung

Lösungen von Salzen in molekularen Lösungsmitteln (z.B. Wasser) zeigen starke Abweichungen vom idealen Verhalten in Bezug auf Dampf-druck- und Schmelzpunktserniedrigung. Die übliche Kryoskopie und Ebullioskopie ist deshalb zur Untersuchung des ionischen Aufbaus von Salzen kaum je brauchbar. Einfachere Verhältnisse sind zu erwarten für *Lösungen von Salzen in salzartigen Medien.* Das führte zur Entwicklung der Salzkryoskopie, bei der ein Zweikomponentensystem — Inertsalz/ Wasser — als Medium dient, mit Fixpunkt ohne Freiheitsgrad (bei dem zwei Festkörper, entweder Eis und Salz oder zwei verschiedene Salzhydrate, sowie die Lösung und die Dampfphase anwesend sind). Die zu untersuchende Verbindung (Fremdsalz) wird als dritte Komponente hinzugefügt und bewirkt eine Temperaturerniedrigung des Fixpunktes.

Die Zugabe des Fremdsalzes bewirkt aber nicht nur die Veränderung der Aktivität der Festkörper in Bezug auf die Lösung, sondern auch diejenige des Wassers, d.h. dessen Dampfdruck. Dieser Effekt ist bisher noch nie für die Untersuchung des Aufbaus von Salzen verwendet worden. Wenn man das unternimmt, kann man von Salzebullioskopie reden, welche analog wie die Salzkryoskopie auf einem Grenzgesetz basiert: Zwei gesättigte Inertsalzlösungen mit festem Inertsalz als Bodenkörper, von denen die eine noch etwas Fremdstoff enthält, besitzen nur dann den gleichen Dampfdruck, wenn die Lösung mit dem Fremdstoff eine um ΔT höhere Temperatur besitzt. Ausgewertet wird das Grenzgesetz, nämlich die thermodynamische Notwendigkeit, daß ΔT zur molalen Konzentration der Fremdpartikeln proportional sein muß, wenn die Fremdsalzkonzentration gegen Null geht. Nur diejenigen Bausteine des Fremdsalzes sind Fremdpartikeln, die im Medium Inertsalz/Wasser nicht bereits enthalten sind und deren Anzahl pro Formelgewicht Fremdsalz (ν_0) gilt es zu bestimmen.

Die hier beschriebene Technik ermöglicht es, die Salzebullioskopie mit kleinen Substanzmengen auszuführen, unter Verwendung des von Hill angegebenen Prinzips: Lösung und Lösungsmittel werden in der gemeinsamen Atmosphäre des Lösungsmittels sich selbst überlassen, wobei die erste eine höhere Temperatur annimmt. Gemessen wird dann nicht der thermodynamisch definierte Temperaturunterschied ΔT, sondern ein sich ausbildender stationärer Temperaturunterschied ΔT_s. Bei der Salzebullioskopie ersetzt man das Lösungsmittel durch die gesättigte Lösung des Inertsalzes. Dieses Medium mit und ohne Fremdsalz (Molalität m_2) bringt man in der gemeinsamen Wasserdampfatmosphäre unter, wobei sich wieder ein stationärer Temperaturunterschied ΔT_s einstellt. Bei rd. 40 untersuchten Systemen zeigte sich, daß folgende lineare Beziehung besteht:

$$\frac{\Delta T_s}{m_2 \cdot K} \equiv \nu(m_2) = \nu_0 + N \cdot m_2 \qquad (1)$$

wobei K eine Eichkonstante bedeutet. Wegen der Linearität ist es nun leicht auf $m_2 \rightarrow 0$ zu extrapolieren und ν_0 zu erhalten, d.h. die Anzahl der Fremdteilchen pro Formel Fremdsalz. Wie hier gezeigt worden ist, kann auch eine Aussage gemacht werden über die Neigung der Geraden

(1), denn N wird zum größten Teil durch einen stöchiometrischen Beitrag N* bedingt und erst in zweiter Linie von der Aktivitätsänderung β_{22}, welche das Fremdsalz erfährt, wenn sich seine Molalität m_2 ändert. Der Aktivitätskoeffizient des Fremdsalzes γ_2 hängt mit (N—N*) zusammen und folgt nicht wie bei der klassischen Ebullioskopie aus dem Verhältnis $\nu(m_2)/\nu_0$.

Wenn man die Molalität des Fremdsalzes auf das Zentralatom bezieht, was besonders bei Metallkomplexen zweckmäßig ist, so bedeutet $1/\nu_0$ die Nuclearität, welche bei Polynuclearen größer als 1 ist. Die Größe ν_0 kann man salzebullioskopisch mit einer Genauigkeit von etwa $\pm 0,05$ bestimmen.

Um zu beweisen, daß auch die nach dem Hill'schen Prinzip — bei dem nicht im thermodynamischen Gleichgewicht gemessen wird — erhaltenen Werte $\nu(m_2)$ und nicht nur ν_0 wahre Kenngrößen sind, wurden auch Dampfdruckmessungen nach der isopiestischen Methode ausgeführt. Man kann auch diese Resultate mit einer linearen Gleichung darstellen:

$$\frac{(p^0-p)\cdot 55,51}{m_2 \cdot p^0} \equiv \nu_p(m_2) = \nu_p^0 + N_p \cdot m_2 \tag{2}$$

wobei p^0 der Partialdruck des Wassers über dem Medium (Inertsalz und dessen gesättigte Lösung) und p derjenige des Mediums mit Fremdsalz der Molalität m_2 bedeutet. Die Zahl 55,51 ist die Molalität des reinen Wassers.

6. Literaturverzeichnis

1. *Löwenherz, L.*: Z. Physik. Chem. *18*, 71 (1895).
2. *Magyar, B.*: Helv. *48*, 1295 (1965).
3. *Hill, A. V.*: Proc. Roy. Soc. (London) A *127*, 9 (1930).
4. *Magyar, B.*: Helv. *51*, 193 (1968).
5. *Haase, R.*: Thermodynamik der Mischphasen. Berlin–Göttingen–Heidelberg: Springer 1956.
6. *Carr, D. S., Harris, B. L.*: Ind. Eng. Chem. *1949*, 2015.
7. *Simon, W., Tomlinson, C.*: Chimia (Aarau) *14*, 301 (1960).
8. *Wegman, D., Tomlinson, C., Simon, W.*: Microchem. J. (Symposium Series) *2*, 1069 (1962).
9. *Higuchi, W. I., Schwartz, M. A.*: J. Phys. Chem. *63*, 996 (1959).
10. *Iyengar, B. R. Y.*: J. Sci. Ind. Res. (India) *13* B, 73 (1954).
11. *Lindner, A.*: Statistische Methoden. Basel: Birkhäuser Verlag 1960.
12. *Schwarzenbach, G., Parissakis, G.*: Helv. *41*, 2425 (1958).
13. *Tobias, R. S.*: J. Inorg. Nucl. Chem. *19*, 348 (1961).

14. *Jahr, K. F., Wegener, K.:* Z. Physik. Chem. (Neue Folge) *53*, 87 (1967).
15. *Strickler, P.:* Chimia (Aarau) *23*, 192 (1969).
16. *Schwarzenbach, G., Magyar, B.:* Helv. *45*, 1429 (1962).
17. *Anderegg, G., Malik, S.:* Chimia (Aarau) *21*, 541 (1967).
18. International Critical Tables *IV*, 254 (1928).
19. *Bousfield, W. R.:* Trans. Faraday Soc. *13*, (1918) 401.
20. *Robinson, R. A., Stokes, R. H.:* Electrolyte Solutions. London: Butterworths 1965.
21. *Scatchard, G., Hamer, W. J., Wood, S. E.:* J. Am. Chem. Soc. *60*, 3061 (1938).
22. *Magyar, B.:* Habilitationsschrift E. T. H. Zürich (1972).

Eingegangen am 1. März 1972

Strukturchemie der Azide

Ulrich Müller

Fachbereich Chemie der Universität Marburg, BRD

Inhalt

1. Summary

Azides can be classified into three groups, depending on their structures: 1. ionic azides, 2. coördinative azides, 3. molecular azides. The main criteria for the assignment of an azide to a certain group are the interatomic distances between its nitrogen atoms and other atoms of the compound. In an *ionic azide* none of these distances is less than the sum of the corresponding ionic radii. In a *coördinative azide* each azido group is coördinated to several metal atoms at distances lying between those for an ionic (or van der Waals) contact and a covalent bond. In *molecular azides* (often called covalent azides) the azido groups are covalently bonded constituents of a molecule (or ion).

Ionic azides are built up from cations and N_3^- ions. These are linear and symmetrical and their N—N bond lengths never deviate significantly from 1.17 Å. The azide ion possesses two delocalized bonding π orbitals (fig. 1a) and can be represented by the resonance formula

$$|: \overset{\ominus}{N} = \overset{\oplus}{N} - \overset{\ominus}{N} :| \quad \longleftrightarrow \quad |: \overset{\ominus}{N} - \overset{\oplus}{N} = \overset{\ominus}{N} :|$$

in which the displaced double bond lines indicate bonding π electrons lying perpendicular to the other bonding π electrons which lie in the plane of the paper; free electron pairs are shown as single lines, non-bonding π electrons as points. Due to resonance stabilization and high lattice energies, ionic azides are much more stable than other azides. The crystal structures of many ionic azides consist of N_3^- and metal ion layers; there are several ways of packing the N_3^- ions in a layer (Fig. 6). Usually the structures resemble those of the corresponding chlorides. LiN_3 and NaN_3 crystallize in distorted NaCl-type structures (Figs. 2 and 3), NH_4N_3 and the isostructural compounds KN_3, RbN_3, CsN_3 and TlN_3 in distorted CsCl-type structures (Fig. 4). As in KN_3, the metal ions of $Ca(N_3)_2$ and $Sr(N_3)_2$ are surrounded by eight azide ions in a nearly quadratic antiprismatic arrangement (Fig. 5), whereas the coördination number of Ba^{2+} in $Ba(N_3)_2$ is nine (Fig. 7). The known essential structural data for ionic azides are summarized in Table 1.

Coördinative azides occur among the heavy metal azides. They must be grouped as intermediates between ionic and molecular azides. They consist of high-polymer crystal networks in which the metal atoms are linked one with another via azido groups. The azido groups are linear and may be symmetrical or not, depending on what surrounds them. Whereas AgN_3 shows a relationship to the ionic azides by crystallizing in a variety of the KN_3 lattice, $Cd(N_3)_2(pyridine)_2$ can almost be classified as a molecular azide due to rather short Cd—N distances. A number of the azides of divalent copper consist of endless chains held together by (shorter) covalent and (longer) coördinative Cu—N bonds (Figs. 8—10). Four different kinds of azido groups with metal-nitrogen distances from 2.58 up to 2.90 Å are found in $Pb(N_3)_2$ (Fig. 11). Structural data for coördinative azides are summarized in Table 2.

Molecular azides are formed with non-metals and transition metals. Usually only one terminal nitrogen atom (αN atom) of the azido group is bonded to one or two other atoms; the azido group is then asymmetrical with bond lengths about 1.25 for αN—βN and 1.12 Å for βN—γN (cf. Fig. 12). These azides possess one delocalized and one localized π bond (Fig. 1b) and can be represented by the resonance formula

$$\underset{R}{\diagup} N \text{-} \overset{\oplus}{N} = \overset{\ominus}{N} : | \quad \longleftrightarrow \quad \underset{R}{\diagup} : \overset{\ominus}{N} \text{---} \overset{\oplus}{N} \equiv N \, | \; .$$

In azides of carbon having π bond systems next to the azido group, the N_3 group is slightly bent and inclined against the plane of the π system by angles up to 20° (Fig. 13). Among these azides the ion $[C(N_3)_3]^+$ is also remarkable in that it is the only known azide in which the N—N bonds are so different in length that they must be regarded as single and triple bonds respectively. In the compounds

$$(BCl_2N_3)_3, \ (SbCl_4N_3)_2, \ (TaCl_4N_3)_2, \ [(N_3)_2 \, Pd(N_3)_2 \, Pd(N_3)_2]^{2-},$$
$$[(CO)_3Mn(N_3)_3Mn(CO)_3]^-$$

αN atoms link two other atoms each, forming four- or six-membered rings (Figs. 14 and 15). Among the transition metal azides the structures of a number of complexes are known, most of which also possess amine or phosphine ligands (Figs. 16 and 17). In molecular azides bonding via both terminal N atoms is less common; it is known for $[Cu(P(C_6H_5)_3)_2 N_3]_2$ in which all N—N bonds have essentially the same length and where the azido groups have an allene-like constitution (Fig. 18). Known structural data for molecular azides are summarized in Table 3.

2. Einleitung

Azide sind energiereiche Verbindungen, die es ermöglichen, unter milden Reaktionsbedingungen Stickstoffverbindungen herzustellen, welche wegen thermischer Instabilität oder sonstiger Umstände auf anderem Weg nicht oder nur mühsam darstellbar sind. Seit den grundlegenden Arbeiten von *Curtius* haben die Azide daher für die präparative Chemie zunehmend an Bedeutung gewonnen. Hierüber sind in den vergangenen Jahren mehrere zusammenfassende Darstellungen erschienen (*1—9*), von denen eine (*9*) auch einen Überblick über einige bindungstheoretische Aspekte der Azide gibt. Weitere Übersichten (*10—13*) befassen sich mit physikalisch-chemischen Aspekten und mit den explosiven Eigenschaften der anorganischen Azide und geben zum Teil auch einen knappen (und nicht immer fehlerfreien) Überblick über die Strukturen einiger ihrer Vertreter. Seit deren Erscheinen sind die Strukturen von einer ganzen Reihe von Aziden neu bestimmt worden, und es ist jetzt möglich, die Azide nach strukturchemischen Gesichtspunkten einzuteilen; dies geschieht im nachfolgenden unter Berücksichtigung der bis Ende 1971 durch Beugungsmethoden oder Mikrowellenspektroskopie ermittelten Azidstrukturen. Vergleiche mit ähnlichen Strukturen anderer Verbindungen werden nicht angestellt, da dies für einige der wichtigsten Strukturtypen der Azide bereits geschehen ist (*14*).

3. Einteilung der Azide

Die Azide lassen sich nach ihrem strukturellen Aufbau in drei Gruppen einteilen:

1. Ionische Azide,
2. koordinative Azide,
3. molekulare Azide.

Unter ionischen Aziden sollen solche verstanden werden, bei denen die Azidogruppe selbst als Ion N_3^- vorkommt; ionische Verbindungen, in welchen die N_3-Gruppe nicht als selbständiges Ion auftritt, sollen nicht hierzu gerechnet werden. Bei den koordinativen Aziden ist die Azidogruppe an mehrere Metallatome mehr oder weniger schwach kovalent („koordinativ") gebunden und ist mit diesen Bestandteil eines hochpolymeren Kristallnetzwerkes; koordinative Azide können nur als Festkörper vorkommen. In den molekularen Aziden, häufig auch als „kovalente Azide" bezeichnet, ist die Azidogruppe ein kovalent gebundener, eindeutiger Bestandteil eines Moleküls (oder Ions).

Nicht alle Azide lassen sich eindeutig einer der drei Gruppen zuordnen, da sich insbesondere die Gruppe der koordinativen Azide nicht scharf gegen die anderen beiden Gruppen abgrenzen läßt. Verbindungen, deren Azidogruppen verschiedenartige Bindungen betätigen, können gleichzeitig zwei der Gruppen angehören.

Als wichtigstes Kriterium für die Zuordnung eines Azids zu einer der drei Gruppen dienen die Abstände d zwischen Atomen der Azidogruppe und anderen Atomen der Verbindung. Der Ionen- und der van-der-Waals-Radius r_i eines N-Atoms einer Azidogruppe liegt je nach den Packungsverhältnissen im Gitter zwischen $r_i = 1,55$ und $r_i = 1,65$ Å, der Kovalenzradius r_c bei 0,70 Å. Ist r_k der Ionenradius eines Kations, so liegt ein ionisches Azid vor, wenn $d \geq r_i + r_k$ für alle Abstände d gilt. Ist r_a der Kovalenzradius eines Atoms, so liegt ein molekulares Azid vor, wenn wenigstens ein Abstand $d = r_c + r_a$ beträgt. Ein koordinatives Azid liegt vor, wenn die Abstände d zwischen $r_c + r_a$ und $r_i + r_k$ (oder r_i plus van-der-Waals-Radius) liegen. Vor allem bei den Schwermetallen sollten die üblicherweise tabellierten Ionenradien allerdings mit Vorsicht verwendet werden, denn sie sind mit einiger Unsicherheit behaftet, weil es von diesen Elementen keine rein elektrostatischen Ionengitter gibt; hier ist es oft vorteilhafter, mit dem van-der-Waals-Radius zu rechnen, dessen Betrag nach *Pauling* etwa $r_a + 0,8$ Å beträgt (17). Für die erste Periode der Übergangsmetalle kann als Faustregel gelten, daß der Metall-Stickstoff-Abstand bei kovalenter Bindung etwa 2,0 Å beträgt, während der van-der-Waals-Abstand bei 3,6 Å liegt; für die zweite und dritte Periode der Übergangsmetalle liegen diese Werte 0,1 bis 0,2 Å höher.

4. Bindungsverhältnisse in den Aziden

Das Azidion ist linear und symmetrisch (Punktsymmetrie $D_{\infty h}$). In allen mit hinreichender Genauigkeit untersuchten Fällen wurden N—N-Abstände gefunden, die kaum von 1,17 Å abweichen. Dieser Wert liegt etwas unter den 1,24 Å für eine N—N-Doppelbindung mit sp^2-hybridisierten N-Atomen (wie z.B. in Azomethan). Da die Bindungsabstände auch von der Hybridisierung der an den σ-Bindungen beteiligten Atomorbitalen abhängen, kann der N—N-Abstand im Azidion trotzdem als Doppelbindung angesprochen werden (im N_3^--Ion ist das mittlere N-Atom sp-hybridisiert, die äußeren N-Atome sind wahrscheinlich nicht hybridisiert; siehe unten).

MO-Betrachtungen führen zu folgendem Bild für das Azidion (9, 18): Außer den beiden σ-Bindungsorbitalen zwischen den N-Atomen sind zwei bindende π-Orbitale besetzt, welche sich über die ganze Länge des Ions erstrecken („delokalisierte π-Bindungen") (Abb. 1a). Außerdem ist an jedem der endständigen Atome ein nicht bindendes freies Elektronenpaar vorhanden und zwei weitere Elektronenpaare befinden sich in nicht bindenden π-Orbitalen. Das mittlere N-Atom trägt eine Formalladung von + 0,6, die äußeren Atome je —0,8. Die Bindungsverhältnisse lassen sich als Mesomerie mit folgenden VB-Grenzformeln wiedergeben:

$$[\,|:\overset{\ominus}{N}=\overset{\oplus}{N}-\overset{\ominus}{N}:|\,\longleftrightarrow\,|:\overset{\ominus}{N}-\overset{\oplus}{N}=\overset{\ominus}{N}:|\,]^-$$

Bei dieser Schreibweise wird die Lage der bindenden π-Elektronen durch die Bindungsstriche derart dargestellt, daß bei übereinanderliegenden Strichen die π-Elektronen in der Papierebene, bei versetzten Strichen senkrecht dazu liegen sollen. Die freien Elektronenpaare werden durch Striche, die nicht bindenden π-Elektronen durch Punktpaare wiedergegeben.

Abb. 1. Schematische Darstellung der bindenden π-Orbitale, a) im Azidion, b) in einer kovalent gebundenen Azidogruppe. Bei letzterer ist auch das freie Elektronenpaar am α-N-Atom gezeigt. Die σ-Bindungen sind als Stäbe dargestellt

Bei den meisten molekularen Aziden betätigt nur eines der endständigen Atome (α-N-Atom genannt) der Azidogruppe eine oder zwei kovalente Bindungen zu anderen Atomen. In diesem Fall ist die Azidogruppe asymmetrisch, weicht aber nur selten von der Linearität ab; der Abstand $\alpha N-\beta N$ liegt im allgemeinen um 1,25 Å, der Abstand $\beta N-\gamma N$ bei 1,12 Å. In der einseitig kovalent gebundenen Azidogruppe liegen zwei verschiedene bindende π-Orbitale vor: eines ist zwischen β-N und γ-N lokalisiert, das andere ist über alle drei N-Atome delokalisiert (Abb. 1b). Nach MO-Berechnungen (9) tragen die Atome folgende Formalladungen: αN: $-0,7$, βN: $+0,9$, γN: $-0,2$. In der VB-Schreibweise läßt sich folgende Mesomerie formulieren:

$$R{\diagup}\,:\overset{\ominus}{N}-\overset{\oplus}{N}\equiv N\,| \quad \longleftrightarrow \quad R{\diagup}N_-\overset{\oplus}{N}=\overset{\ominus}{N}\,:|$$

In seltenen Fällen ist die Azidogruppe mit beiden Endatomen kovalent an je ein anderes Atom gebunden. Sie hat dann eine Struktur wie das Allen mit kumulierten N=N-Doppelbindungen, die genauso wie im N_3^--Ion etwa 1,17 Å lang sind:

$$R{\diagup}N_-N=N\,|\,\blacktriangleright R$$

Koordinative Azide kann man sich aus ionischen Aziden entstanden denken, wenn man von den nächsten Gegenionen im gedachten Ionengitter einige an die Azidogruppe näherrückt, wobei (zunächst schwache) kovalente Wechselwirkungen zustande kommen. Koordinative Azide können dementsprechend als Zwischending zwischen den ionischen und den molekularen Aziden betrachtet werden. Auch in den koordinativen Aziden ist die Azidogruppe praktisch immer linear; ob ihre beiden N—N-Bindungen gleich lang sind oder nicht, hängt von der Umgebung ab. Sind z. B. jedem der endständigen N-Atome zwei andere Atome benachbart, so sind die N—N-Abstände praktisch gleich groß, hat aber eines der endständigen N-Atome einen, das andere drei nächste Nachbarn, so ist der N—N-Abstand zu dem dreifach koordinierten N-Atom länger. Wegen der Wechselwirkung mit den Nachbaratomen ist das π-Elektronensystem der koordinativ gebundenen Azidogruppe weitgehend lokalisiert.

Allen Aziden ist gemeinsam, daß stets nur die endständigen N-Atome der N_3-Gruppe mit anderen Atomen in Wechselwirkung treten; auch bei den ionischen Aziden ist das Kation immer diesen Atomen benachbart, in Übereinstimmung mit der Vorstellung, daß sich die negative Ladung des N_3^--Ions auf diese Atome konzentriert.

Während das mittlere N-Atom in allen Fällen als sp-hybridisiert anzusehen ist, hängt die Hybridisierung der beiden endständigen N-Atome von den Bindungen der Azidogruppe zu umgebenden Atomen ab.

Die Art der Hybridisierung läßt sich an den Bindungswinkeln ablesen. Für das Azidion ist anzunehmen, daß jedes der endständigen N-Atome nicht hybridisiert ist, d.h. das freie Elektronenpaar befindet sich in einem s-Orbital, ein p-Orbital beteiligt sich an der σ-Bindung und die übrigen zwei p-Orbitale gehen in die π-Orbitale ein. Denkbar wäre auch eine sp-Hybridisierung, bei welcher das freie Elektronenpaar ein sp-Orbital besetzen würde; dies widerspricht jedoch der Beobachtung, daß bei molekularen Aziden der Bindungswinkel am α-N-Atom nie 180° beträgt. Wären im N_3^--Ion die endständigen Atome sp-hybridisiert, so müßte es möglich sein, eine kovalente Bindung zu knüpfen, ohne daß sich die Hybridisierung dabei ändert; ist das endständige Atom im N_3^--Ion jedoch nicht hybridisiert, so ist der Übergang zur kovalenten Bindung nur möglich, wenn ein neues Hybrid gebildet wird; dieses ist bevorzugt ein sp^2-Hybrid, denn nur dieses läßt ein p-Orbital für das π-System frei und bedingt gleichzeitig einen Bindungswinkel von 120°. Die tatsächlich beobachteten Winkel am α-N-Atom von molekularen Aziden liegen in der Gegend von 120°, wenn auch eine Tendenz zu kleineren Winkeln (bis 107°) zu verzeichnen ist; dies läßt sich als Folge einer Abstoßung der Bindungen durch das freie Elektronenpaar deuten (16). Bei größeren Molekülen bedingen sterische Behinderungseffekte mitunter auch Winkel, die etwas über 120° liegen. In koordinativen Aziden läßt sich kein einheitliches Bild für die Art der Hybridisierung der endständigen N-Atome erkennen; in einigen Fällen ist die Anordnung der koordinierten Metallatome so, daß man eine Tendenz zur sp^3-Hybridisierung vermuten kann.

Der zunehmenden Lokalisierung der π-Elektronen entsprechend, kann man nach den oben aufgeführten Überlegungen eine Abnahme der Stabilität in folgender Reihenfolge erwarten: N_3^--Ion > einseitig kovalent gebundene N_3-Gruppe > zweiseitig kovalent oder koordinativ gebundene N_3-Gruppe. Dies Bild entspricht qualitativ der Beobachtung, daß koordinative Azide sehr leicht explodieren, molekulare Azide zwar meist explosiv sind, sich aber bei Beachtung gewisser Vorsichtsmaßregeln einigermaßen sicher handhaben lassen, während die ionischen Azide nicht explosiv sind. Trotzdem handelt es sich um ein einseitiges Bild, und insbesondere die sehr viel höhere Stabilität der ionischen Azide wird damit nicht befriedigend erklärt. Bei den Stabilitätsbetrachtungen muß stets auch die Gitterenergie des Kristalls berücksichtigt werden. Diese ist bei den ionischen Aziden so groß, daß die bei der Zersetzung des Azids freiwerdende Energie vom zusammenbrechenden Kristallgitter aufgezehrt wird, und es kann nicht zu einer thermischen Explosion kommen. Die Werte für die Gitterenergie der (ionischen) Alkaliazide liegen zwischen 150 und 200 kcal/Mol, für die Azide der schweren Erdalkalimetalle zwischen 450 und 520 kcal/Mol (13, 15, 72). Für die (koordinativen) Azide der Schwermetalle werden formal ebenfalls Werte in dieser Größenordnung berechnet, doch liegt diesen Berechnungen die falsche Annahme eines rein elektrostatischen Ionengitters zugrunde, und die tatsächlichen Gitterenergien dürften beträchtlich niedriger sein.

Die Stabilität der Azide kommt quantitativ in ihren Bildungsenthalpien zum Ausdruck. Für die ionischen Azide liegen sie um Null kcal/Mol, für die sehr explosiven Schwermetallazide über $+100$ kcal/Mol (13, 15, 74). Da die Höhe der Ionisierungsenergie eines Metalls auf den Ionencharakter seiner Verbindungen von entscheidendem Einfluß ist und auch in die Gitterenergie mit eingeht, besteht bei den Metallaziden ein direkter Zusammenhang zwischen Stabilität und Ionisierungsenergie (11); je kleiner die Ionisierungsenergie, desto stabiler ist das Azid. Die Bildungsenthalpie des gasförmigen N_3^--Ions wurde zu 34, die des gasförmigen N_3-Radikals zu 104 kcal/Mol berechnet (72).

5. Ionische Azide

Typische Vertreter der ionischen Azide sind die Azide der Alkali- und und der schweren Erdalkalimetalle sowie einige Azide mit komplexen Kationen, z. B. Tetraalkylammoniumazide.

Ionische Azide zeichnen sich durch ihre chemische Stabilität aus. Sie zersetzen sich erst bei erhöhten Temperaturen in kontrollierbarer Weise und sind gegen mechanische Beanspruchung unempfindlich. Sie sind leicht löslich in stark polaren Lösungsmitteln, insbesondere in Wasser, wenig löslich in schwächer polaren Lösungsmitteln, z. B. in flüssigem SO_2, und unlöslich in unpolaren Lösungsmitteln.

Präparativ hat vor allem das Natriumazid Bedeutung als Ausgangsmaterial zur Darstellung der meisten anderen Azide; darüber hinaus spielen ionische Azide nur in Sonderfällen eine Rolle für die präparative Chemie.

Die wichtigsten Daten der bekannten Kristallstrukturen von ionischen Aziden sind in Tabelle 1 zusammengestellt. Die meisten Strukturen lassen sich von denen der entsprechenden Chloride ableiten, wenn man die Cl^--Ionen durch N_3^--Ionen ersetzt, bei gleichzeitiger Verzerrung des Gitters um der zylindrischen Gestalt des N_3^--Ions Rechnung zu tragen. Für die Art der Verzerrung werden verschiedene Möglichkeiten realisiert.

Vom Natriumazid sind zwei Modifikationen bekannt, deren Strukturen dem NaCl-Typ ähnlich sind (19). Werden im NaCl die Cl^--Ionen durch N_3^--Ionen so ersetzt, daß diese parallel zur Raumdiagonalen der NaCl-Elementarzelle zu liegen kommen, so kommt man zur Struktur des β-NaN₃. Die Raumdiagonale wird dabei gedehnt und aus der NaCl-Zelle wird eine flächenzentrierte rhomboedrische NaN_3-Elementarzelle mit einem Rhomboederwinkel von $\alpha' = 67,1°$ (Abb. 2). Den kristallographischen Konventionen entsprechend wird als Elementarzelle üblicherweise die primitive Zelle mit $\alpha = 38,7°$ angegeben. Die β-NaN₃-Struktur kann auch als Schichtstruktur beschrieben werden, mit sich abwechselnden Schichten von N_3^-- und Na^+-Ionen parallel zu (111), welche wie in einer kubisch dichtesten Kugelpackung gestapelt sind; die N_3^--Ionen stehen senkrecht zur Schichtebene (Abb. 2, 6a). Unterhalb von 18 °C wandelt sich das β-NaN₃ in eine monokline α-Modifikation um. Bei der Umwandlung werden die Schichten in Richtung senkrecht zur b-Achse scherenartig um 4,8° gegeneinander versetzt; die N_3^--Ionen neigen sich dabei um 12° gegen die ursprüngliche Raumdiagonale (Abb. 3) (19).

Isostrukturell zum α-NaN₃ kristallisiert auch das LiN₃ (19).

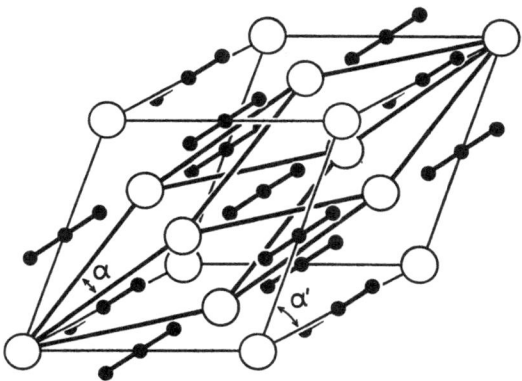

Abb. 2. Rhomboedrische Elementarzellen von β-NaN₃. Die äußere Zelle mit dem Rhomboederwinkel $\alpha' = 67,1°$ ist eine flächenzentrierte Zelle vom NaCl-Typ; die innere, stark umrandete Zelle mit $\alpha = 38,7°$ entspricht der Aufstellung nach den kristallographischen Konventionen. Man erkennt die schichtenweise Anordnung der Atome. Große Kreise = Na^+-Ionen, Hanteln = N_3^--Ionen

KN₃, RbN₃, CsN₃ und TlN₃ kristallisieren in einer tetragonal verzerrten Variante des CsCl-Typs und können ebenfalls als Schichtenstrukturen beschrieben werden (20). Die Schichten liegen senkrecht zur tetragonalen c-Achse, und folgen derart aufeinander, daß alle Metallionen sich decken, während jede Azidschicht sich immer nur mit der jeweils übernächsten Schicht deckt (Abb. 4). Anders als beim NaN₃ liegen die Azidionen in der Schicht und haben darin zwei zueinander senkrechte Orientierungen. Die Stapelfolge der Schichten ist so, daß die acht nächsten N-Atome um ein Metallion die Eckpunkte eines wenig

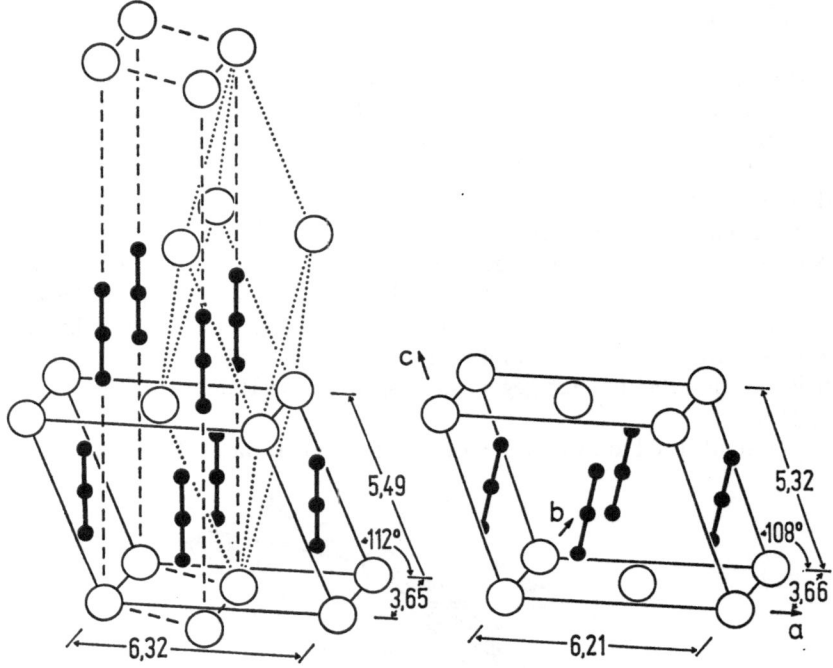

Abb. 3. Elementarzellen von β-NaN$_3$ (links) und α-NaN$_3$ (rechts) zum Vergleich. Beim β-NaN$_3$ werden drei Aufstellungen für die Elementarzelle gezeigt: punktiert: rhomboedrische Aufstellung; gestrichelt: trigonale Aufstellung; ausgezogen: basiszentrierte monokline Aufstellung. Letztere zeigt die Verwandtschaft zur monoklinen Elementarzelle des α-NaN$_3$

verzerrten quadratischen Antiprismas einnehmen (Antiprisma mit quadratischen Deckflächen, die 38 (CsN$_3$) bis 41° (KN$_3$) gegeneinander verdreht sind). RbN$_3$ und CsN$_3$ wandeln sich bei 315 °C bzw. 151 °C in eine kubische Modifikation vom CsCl-Typ um, bei welcher eine statistische Orientierung der N$_3^-$-Ionen angenommen wird (21).

Ammoniumazid kristallisiert in einer orthorhombisch verzerrrten CsCl-Anordnung, die große Ähnlichkeit zur Struktur von Kaliumazid hat (22, 23). Werden alle Azidschichten des KN$_3$ deckungsgleich gestapelt und wird die Hälfte der Azidionen jeder Schicht um 23° gegen die Schichtebene verdreht, so werden von den acht Azidnachbarn jeweils vier N-Atome in tetraedrischer Anordnung näher um das Kation gerückt; diese vier Atome betätigen Wasserstoffbrückenbindungen mit dem NH$_4^+$-Ion. Wegen der Existenz von Wasserstoffbrücken bestehen gerichtete Bindungskräfte, so daß das NH$_4$N$_3$ den koordinativen Aziden nahesteht.

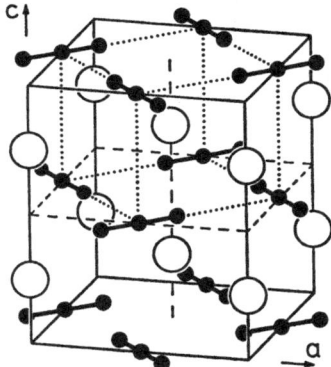

Abb. 4. Tetragonale Elementarzelle von KN_3. Punktiert: Pseudozelle vom CsCl-Typ. Wird durch Verkürzung der a-Achse diese Zelle orthorhombisch verzerrt, so kommt man zur Elementarzelle des AgN_3

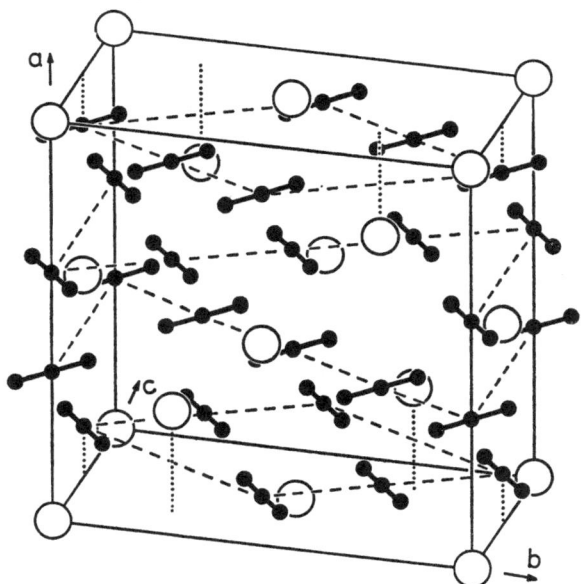

Abb. 5. Orthorhombische Elementarzelle von $Sr(N_3)_2$. Gestrichelt ist angedeutet, welche N_3^--Ionen jeweils in einer der vier verschiedenen Schichten liegen

Eine gewisse Ähnlichkeit zum Strukturtyp des Kaliumazids zeigt auch die Struktur des Strontiumazids (*19*). Hier liegen nahezu planare Schichten von in der Schichtebene liegenden N_3^--Ionen vor, zwischen denen sich die Sr^{++}-Ionen in einer verzerrt quadratisch-antiprismatischen Umgebung von acht nächsten N-Atomen befinden (Quadrate des Antiprismas 35° gegeneinander verdreht). Die orthorhombische Elementarzelle enthält von jeder der beiden Ionensorten vier verschiedene Schichten senkrecht zur a-Achse (Abb. 5). Dem Calciumazid wird dieselbe Struktur zugesprochen (24).

Für die bisher besprochenen Strukturen der Azide von Li, Na, K, Rb, Cs, Tl, NH_4 und Sr ist die schichtenweise Packung der Azidionen charakteristisch. Die unterschiedlichen Arten der Anordnung von N_3^--Ionen in einer Schicht wird in Abb. 6 veranschaulicht. Geht man davon aus, daß die negative Ladung des Ions sich auf dessen endständige Atome konzentriert, während das mittlere N-Atom eine partiell positive Ladung trägt (siehe Abschn. 4), so muß aus elektrostatischen Gründen eine Ionenpackung bevorzugt sein, bei welcher die Azid-Endatome untereinander möglichst weit entfernt sind, dagegen den Mittelatomen möglichst nahe kommen. Solch eine Anordnung ist in den Schichten des KN_3-Typs ideal verwirklicht (Abb. 6e); beim KN_3 selbst beträgt der kürzeste N...N-Abstand zwischen benachbarten N_3^--Ionen 3,14 Å. Ein etwas größerer Abstand wird beim NH_4N_3 gefunden, dessen Azidschichten ebenso günstig gepackt sind (Abb. 6c). Die stärkeren elektrostatischen Wechselwirkungen zwischen Azid und den kleineren Ionen Li^+ und Na^+ sowie den doppelt geladenen Ionen Ca^{++} und Sr^{++} führen aber zu Anordnungen, die bei Betrachtung der Packung der N_3^--Ionen alleine weniger günstig erscheinen. Dies trifft vor allem für das β-NaN_3 zu, dessen N_3^--Ionen genau aneinander liegen und dementsprechend den relativ großen N...N-Abstand von 3,65 Å voneinander haben (Abb. 6a). Die Verdrehung der Azidionen gegen die Schichtebene um 12° in der Tieftemperaturmodifikation α-NaN_3 bedingt, daß von den drei nächsten N_3^--Nachbarn eines Azid-Ions nur noch eines ihm genau gegenüber liegt und im Abstand von 3,65 Å verbleibt, während die anderen beiden in eine etwas günstigere Lage verschoben sind und in einer Entfernung von 3.54 Å liegen (Abb. 6b).

Die Lage der N_3^--Ionen in einer Schicht des $Sr(N_3)_2$ läßt sich insofern mit der im α-NaN_3 vergleichen, als alle Ionen parallel liegen und jedes N_3^--Ion mit der Richtung zum Nachbar-Ion einen Winkel von 12° einschließt und 3.17 Å davon entfernt ist; anders ist hingegen, daß die N_3^--Ionen in der Schicht liegen (Abb. 6d).

In allen Fällen sind die Azidschichten so gestapelt, daß die Kationen sich gegenüber von Lücken in der Schicht befinden; die Schichten liegen dadurch nahe beieinander und berühren sich in den meisten Fällen. In einigen Fällen sind die N...N-Abstände innerhalb einer Schicht sogar größer als die zwischen zwei Schichten.

Eine andere Art von Schichtenaufbau weist die Struktur von Bariumazid auf (25, 26). Hier liegen die Ba^{++}- und die N_3^--Ionen in derselben

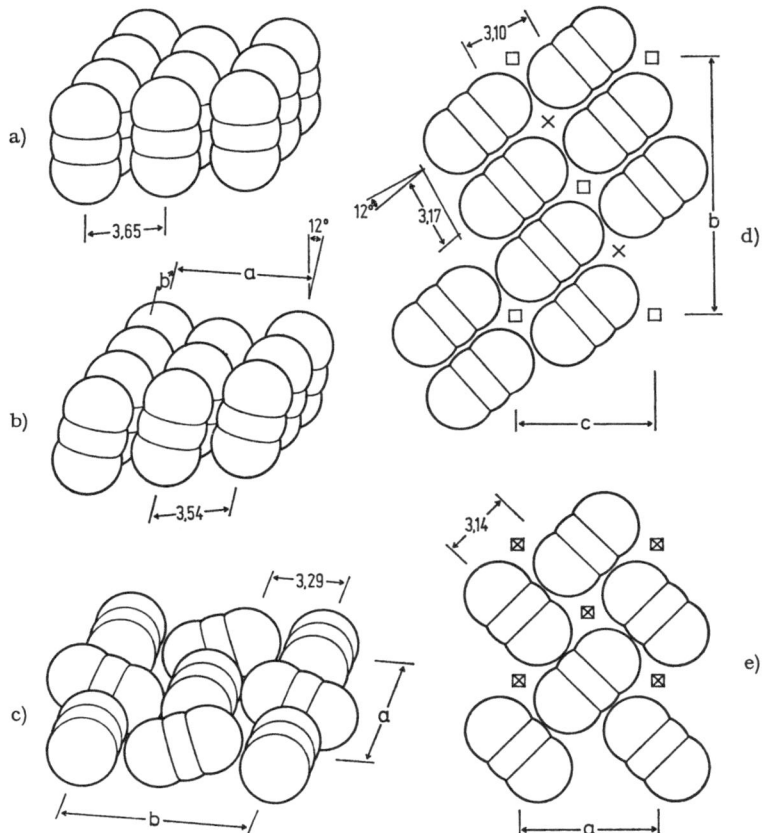

Abb. 6. Verschiedene Arten der Packung von N₃⁻-Ionen in einer Schicht. a) Im β-NaN₃. b) Im α-NaN₃. c) Im NH₄N₃. d) Im Sr(N₃)₂. e) im KN₃. Bei d) und e) wird angedeutet, wo sich die benachbarten Metallionen über (kleine Quadrate) und unter (Kreuze) der Schicht befinden

planaren Schicht und zwei um 180° gegeneinander verdrehte Schichten sind abwechselnd aufeinander gestapelt. (Abb. 7). Die Struktur hat dasselbe Aufbauprinzip wie die von BaCl₂. Das Koordinationspolyeder um ein Ba⁺⁺-Ion ist ein trigonales Prisma mit aufgesetzten Pyramiden über den rechteckigen Flächen, so daß jedes Ba⁺⁺ von neun N-Atomen aus neun Azidionen umgeben ist. Es liegen zwei kristallographisch verschiedene Azidionen vor, von denen eines von vier, das andere von fünf Ba⁺⁺-Ionen umgeben ist. Im Gegensatz zum vierfach koordinierten N₃⁻ ist das fünffach koordinierte N₃⁻-Ion nicht exakt symmetrisch; dies

Tabelle 1. *Übersicht über die wichtigsten Strukturdaten von ionischen Aziden*

Wenn einzelne Parameter mehrmals bestimmt wurden, so sind nur die neuesten oder die genauesten Werte aufgeführt. Alle Abstände sind in Å, Winkel in Grad angegeben. Die Summe der Ionenradien wurde mit dem Wert $r_N = 1,6$ Å für Stickstoff berechnet. Bestimmungsmethoden: X = Röntgen-, N = Neutronenbeugung. R ist der kristallographische Zuverlässigkeitsindex

Verbindung	Elementarzelle	Raum-gruppe	For-mel-einh. pro El.-zelle	N-N-	kürzester N...N- Abstand	N- Schicht- Abstand	Anzahl M+ um ein End-N, Abstand	Sum-me Ionen-radien	Kationen-koordina-tion	Best.-meth.	R(%)	Li-te-ra-tur
LiN_3	$a=5,627\ b=3,319$ $c=4,979\ \beta=107,4$	C2/m	2	1,162	3,100	2,47	$\begin{cases}2\times2,29\\1\times2,21\end{cases}$	2,2	6 ca. oktaedr.	X	12	(19)
$LiN_3 \cdot H_2O$	$a=9,259\ c=5,594$	P6$_3$/mcm	6	1,174			$2\times2,226$	2,2	6 oktaedr.	X	4,2	(27)
$\alpha\text{-}NaN_3$	$a=6,211\ b=3,658$ $c=5,323\ \beta=108,4$	C2/m	2	1,167	3,45	2,769	$\begin{cases}2\times2,54\\1\times2,44\end{cases}$	2,5	6 ca. oktaedr.	X	8	(19)
$\beta\text{-}NaN_3$	$a=5,491\ \alpha=38,7$	R$\bar{3}$m	1	1,173	3,443	2,725	$3\times2,507$	2,5	6 oktaedr.	X	12,5	(19)
NH_4N_3	$a=8,930\ b=8,642$ $c=3,800$	Pman	4	1,16 1,17	3,29	3,80	$2\times2,94$ (N–H··N)		4 tetraedr. H-Brücken	X	?	(22)
KN_3	$a=6,113\ c=7,094$	I4/mcm	4	1,180	3,142	3,547	$4\times2,963$	2,9	8 ca. quadr. antiprism.	X	3,8	(20)
RbN_3	$a=6,310\ c=7,519$	I4/mcm	4	1,17	3,29	3,76	$4\times3,10$	3,1	8 ca. quadr. antiprism.	X	9,8	(20)
CsN_3	$a=6,541\ c=8,091$	I4/mcm	4	1,15	3,47	4,05	$4\times3,28$	3,3	8 ca. quadr. antiprism.	X	5,6	(20)

TlN₃	$a=6{,}208\ c=7{,}355$	I4/mcm	4	1,16	3,23	3,68	$4\times3{,}04$	3,0	8 ca. quadr. antiprism.	X	7,0	(20)
Ca(N₃)₂	$a=11{,}32\ b=11{,}07$ $c=5{,}95$	Fddd	8	?	?	?	$2\times$	2,6	8 ca. quadr. antiprism.	X	—	(24)
Ca(N₃)₂· ½ H₂O	$a=13{,}61\ b=6{,}17$	?	3?	?	?	?	?	2,6	?	X	—	(24)
Ca(N₃)₂· 1½ H₂O	$a=11{,}59\ b=6{,}14$ $c=7{,}83\ \beta=106{,}7$	?	4	?	?	?	?	2,6	?	X	—	(24)
Ca(N₃)₂· 4 H₂O	$a=11{,}15\ b=6{,}27$ $c=10{,}18\ \beta=107{,}7$	?	?	?	?	?	?	2,6	?	X	—	(24)
Sr(N₃)₂	$a=11{,}82\ b=11{,}47$ $c=6{,}08$	Fddd	8	1,16	3,10	2,96	$\begin{cases}1\times2{,}60\\1\times2{,}70\end{cases}$	2,7	8 ca. quadr. antiprism.	X	9,5	(19)
Ba(N₃)₂	$a=9{,}59\ b=4{,}39$ $c=5{,}42\ \beta=99{,}7$	P2₁/m	2	$\begin{cases}1{,}168\\1{,}164\end{cases}$ $\begin{cases}1{,}178\\1{,}157\end{cases}$	3,061	(2,71)	$1\times2{,}914+$ $1\times2{,}947$ $2\times2{,}895$ $2\times2{,}887+$ $1\times2{,}985$ $2\times2{,}883$	2,9	9 tetrakaidekaedr.	N	4,1	(25)
Ba(N₃)₂· H₂O	$a=7{,}29\ b=10{,}84$ $c=6{,}96\ \beta=104{,}7$	C2/c (Cc?)	4	?	?	?	?	2,9	?	X	—	(28)
Ba(N₃)₂· 1½ H₂O	$a=7{,}58\ b=5{,}22$ $c=14{,}56\ \beta=93{,}1$	?	?	?	?	?	?	2,9	?	X	—	(71)
[Co(NH₃)₅ N₃](N₃)₂	$a=12{,}997\ b=8{,}031$ $c=10{,}414$	Pnam	4	$\begin{cases}1{,}17\\1{,}16\end{cases}$	(3,05)	—	—	—	CsCl-Typ ähnlich	X	7,5	(63)

kann dadurch erklärt werden, daß das eine endständige N-Atom, welches drei Ba^{++}-Ionen berührt, einen höheren Anteil der negativen Ladung haben muß als das andere, an das nur zwei Ba^{++}-Ionen anliegen.

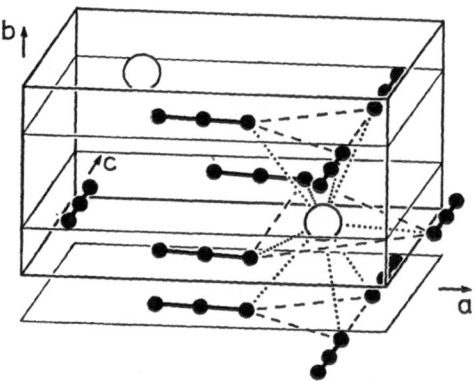

Abb. 7. Elementarzelle von $Ba(N_3)_2$. Die dünn ausgezogenen Linien zeigen an, in welchen Ebenen die Atome liegen. Das Koordinationspolyeder um ein Ba^{++}-Ion ist dargestellt

Von den Hydraten sind nur die Struktur von $LiN_3 \cdot H_2O$ (*27*) und die Elementarzellen mehrerer Hydrate von $Ca(N_3)_2$ und $Ba(N_3)_2$ bekannt (*24, 28, 71*). Im $LiN_3 \cdot H_2O$ liegen zwei kristallographisch verschiedene Li^+-Ionen vor; eines ist von sechs H_2O-Molekülen, das andere von sechs N_3^--Ionen umgeben. Die Wassermoleküle betätigen Wasserstoffbrückenbindungen zu den Azidionen.

6. Koordinative Azide

Zu den koordinativen Aziden zählen in erster Linie Azide der Schwermetalle; ihrer polymeren Struktur entsprechend sind sie schwerlöslich in allen Lösungsmitteln. Sie sind sehr leicht zur Explosion zu bringen und spielen zum Teil eine Rolle in der Sprengstofftechnik. Außer dem Silberazid haben sie praktisch keine Bedeutung für die präparative Chemie.

In Tabelle 2 sind bekannte Kristalldaten von koordinativen Aziden zusammengestellt. Die angegebenen Metall-Stickstoff-Abstände geben einen Eindruck von der Art der Bindungen.

Eine deutliche Verwandtschaft zu den ionischen Aziden zeigt das Silberazid. Es kristallisiert in ähnlicher Weise wie das Kaliumazid; die Elementarzelle ist jedoch orthorhombisch verzerrt, derart, daß von den acht nächsten N-Atomen um das Metallatom vier näher an das Ag gerückt sind (*30, 31*) (vgl. Abb. 4). Diese vier N-Atome umgeben das Ag-Atom in einer verzerrt tetraedrischen Anordnung mit Ag-N-Abständen von 2,56 Å. Die Azidogruppen sind linear und symmetrisch.

Dem CuN_3 wird eine tetragonale Struktur zugeschrieben, welche ein anderes Aufbauprinzip als jede der bekannten Azidstrukturen hat (*32*). Jedes Cu-Atom soll danach in planarer Anordnung von vier nächsten N-Atomen umgeben sein, zwei (in trans-Position) in 2,23, zwei in 2,30 Å Abstand mit N-Cu-N-Winkeln von 73° und 107°. Jedes endständige N-Atom der Azidogruppen ist dann zwei nächsten Cu-Atomen benachbart, von denen eines ungefähr in Richtung der verlängerten Azidogruppe, das andere etwa senkrecht dazu liegt.

Für die Azide des zweiwertigen Kupfers ist charakteristisch, daß das Cu-Atom immer annähernd quadratisch planar von vier kovalent gebundenen Liganden umgeben ist. Zwei weitere Liganden, welche sich über und unter der Ebene des Koordinationsquadrates in den Spitzenpositionen eines verzerrten Oktaeders befinden, sind weiter entfernt und gehören gleichzeitig zu den kovalent gebundenen Liganden zweier weiterer Cu-Atome. Diese Verbindungen können daher gleichzeitig den molekularen und den koordinativen Aziden zugeordnet werden. Im Falle des $Cu(N_3)_2$ liegen unendliche Ketten vor, in denen je zwei Cu-Atome über zwei N-Atome unter Ausbildung eines Vierringes kovalent miteinander verknüpft sind (Abb. 8) (*33, 34*). Jede Kette liegt versetzt parallel neben einer zweiten, wobei die Hälfte der α-N-Atome der einen Kette

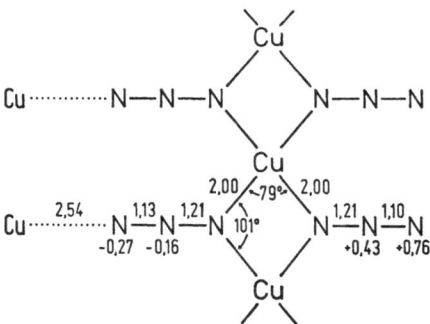

Abb. 8. Ausschnitt aus einer unendlichen Kette im $Cu(N_3)_2$. Abstände in Å. Die Zahlen mit Vorzeichen geben an, wieviel Å das betreffende Atom außerhalb der Kettenebene liegt. Die rechts stehenden α-N-Atome koordinieren Cu-Atome einer darunterliegenden Kette

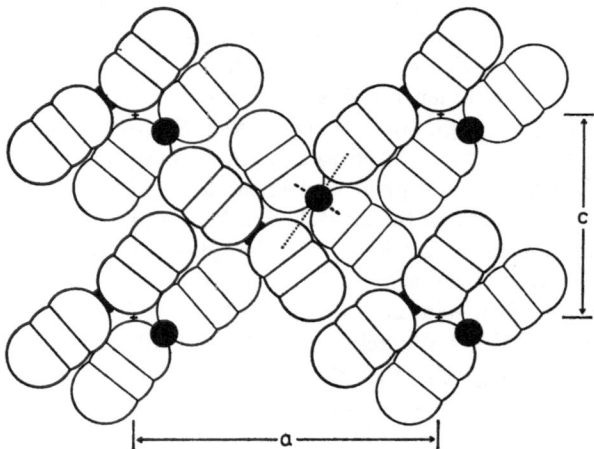

Abb. 9. Packung der Ketten im $Cu(N_3)_2$. Die Ketten verlaufen senkrecht zur Papierebene. Die stärker ausgezogenen N_3-Gruppen liegen eine halbe Kettenperiode über den schwächer ausgezogenen. Gestrichelt: Cu—N-Bindungen in einer Kette; punktiert: Cu · · N-Koordination zwischen den Ketten

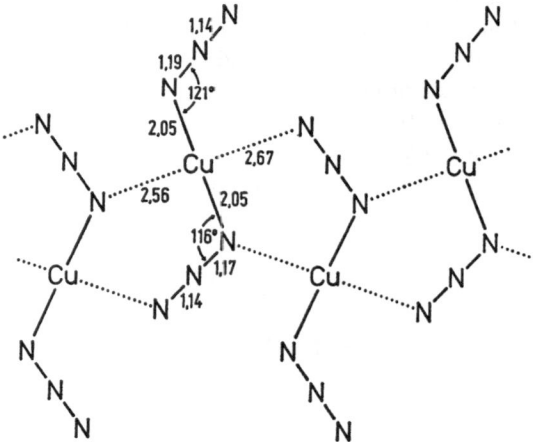

Abb. 10. Ausschnitt aus einer unendlichen Kette im $Cu(NH_3)_2(N_3)_2$. Die Ketten sind planar. Über und unter jedem Cu-Atom befindet sich je eine NH_3-Gruppe im Abstand von 1,99 Å. Völlig analog ist auch $Cu(Pyridin)_2(N_3)_2$ aufgebaut

die Cu-Atome der anderen koordiniert; die beteiligten Azidogruppen sind um 20° aus der Kettenebene herausgewinkelt; die übrigen Azidogruppen koordinieren mit ihren γ-N-Atomen Cu-Atome von weiteren

Kettenpaaren (Abb. 9). Bei den Verbindungen $Cu(N_3)_2(NH_3)_2$ (*35*) und $Cu(N_3)_2(Pyridin)_2$ (*36*) liegen ebenfalls unendliche Ketten vor, die jedoch etwas anders geknüpft sind (Abb. 10). Jedes Cu-Atom ist an zwei N_3-Gruppen und zwei NH_3- bzw. Pyridinreste in trans-Stellung kovalent gebunden und je ein α-N- und ein γ-N-Atom von Azidogruppen anderer Cu-Atome besetzen die Spitzenpositionen des verzerrten Oktaeders. Beim $Cu(N_3)_2$ ist das α-N-Atom der Azidogruppe an zwei Cu-Atome kovalent gebunden, bei den anderen beiden Verbindungen an jeweils eines. In allen Fällen liegen die kovalenten Cu—N-Bindungslängen bei 2,00 Å und die kürzesten koordinativen Cu...N-Abstände bei 2.55 Å. Azidoverbindungen des Kupfers mit molekularem Aufbau siehe Abschnitt 6.

Während vom $Cd(N_3)_2$ nur die Elementarzelle bekannt ist (*37*), wurde die Struktur des $Cd(N_3)_2(Pyridin)_2$ genauer untersucht (*38*). Jedes Cd-Atom ist darin oktaedrisch von zwei Pyridin- und vier Azidogruppen umgeben und jede Azidogruppe wirkt über ihre zwei endständigen N-Atome als Brücke zwischen zwei Cd-Atomen. Die Azidogruppen sind fast symmetrisch und ihre zwei Cd—N-Bindungen liegen so zueinander, daß jede N_3-Gruppe eine dem Allen ähnliche Konfiguration hat. Die Bindungswinkel Cd—N—N haben Werte von 129° und 139°, und die Cd—N-Abstände von 2,34 und 2,35 Å sind nur wenig größer als für eine kovalente Bindung. $Cd(N_3)_2(Pyridin)_2$ kann daher auch als hochpolymeres molekulares Azid angesehen werden. (Wegen der Verknüpfung über beide endständigen N-Atome der Azidogruppe siehe auch die Struktur von $[Cu(P(C_6H_5)_3)_2N_3]_2$, Abschnitt 7.)

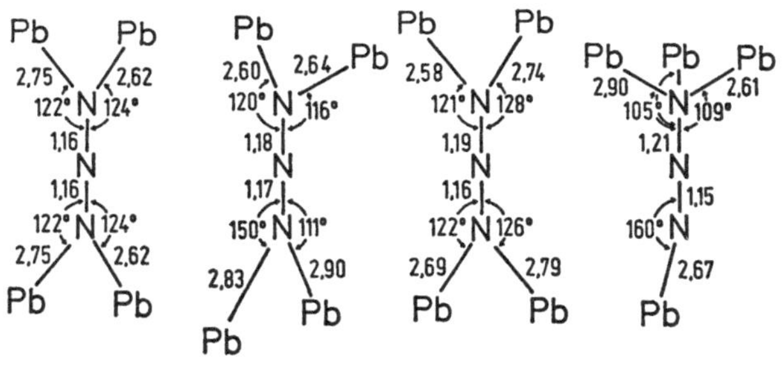

Abb. 11. Aufbau und Umgebung der vier verschiedenen Azidogruppen im α-$Pb(N_3)_2$

Vom $Pb(N_3)_2$ sind vier Modifikationen bekannt, von denen bisher nur die Struktur des α-$Pb(N_3)_2$ aufgeklärt wurde (*39*). In ihr kommen vier verschiedene, lineare Azidogruppen vor, von denen zwei symmetrisch sind, eine wenig asymmetrisch ist, und eine deutlich verschiedene N—N-

Tabelle 2. *Übersicht über die wichtigsten Strukturdaten von koordinativen Aziden*
Es gelten dieselben Angaben wie in der Legende zu Tabelle 1 aufgeführt. Metall-Stickstoff-Abstände zum Vergleich: kovalent 1,9—2,0 Å für Cu—N, 2,0 bis 2,1 Å für Ag—N und 2,2 Å für Pb—N und Cd—N; van-der-Waals-Abstände ca. 3,6 Å. Py = Pyridin

Verbindung	Elementarzelle	Raumgruppe	Formeleinh. pro El.-zelle	N-N-Abstände	Anzahl Metallatome um ein End-N, Abstand	Metallatom-koordination	Best.-meth.	R (%)	Literatur
CuN_3	$a=8,65$ $c=5,59$	$I4_1/a$	8	1,17	$1\times2,23+1\times2,30$	4 planares Parallelogramm	X	?	(32)
$Cu(N_3)_2$	$a=13,454$ $b=3,079$ $c=9,084$	Pnma	4	1,21 1,09 1,21 1,13	$2\times2,00+1\times2,71$ $2\times2,00$ $1\times2,54$	4+2 verzerrt oktaedrisch	X	5,5	(33, 34)
$Cu(N_3)_2(NH_3)_2$	$a=6,389$ $b=7,454$ $c=12,71$	Pnma	4	1 19 1,14 1,17 1,14	$1\times2,05+1\times2,56$ $1\times2,05$ $1\times2,67$	4+2 verzerrt oktaedrisch	X	7,5	(35)
$Cu(N_3)_2Py_2$	$a=13,884$ $b=13,646$ $c=6,406$	$Cmc2_1$	4	1,23 1,19 1,14 1,08	$1\times2,01+1\times2,76$ $1\times1,98$ $1\times2,50$	4+2 verzerrt oktaedrisch	X	7,8	(36)
AgN_3	$a=5,617$ $b=5,915$ $c=6,006$	Ibam	4	1,16 angenommen	$2\times2,56+2\times2,79$	4+4 verz. quadr. antiprismat.	X	?	(30, 31)
$Cd(N_3)_2$	$a=7,82$ $b=6,46$ $c=10,04$	Pbca	8	?	?	?	X	—	(37)
$Cd(N_3)_2Py_2$	$a=15,795$ $c=10,148$	$I4_1/a$	8	1,17 1,14	$1\times2,34$ $1\times2,35$	6 oktaedr.	X	5,6	(38)
$\alpha\text{-}Pb(N_3)_2$	$a=11,31$ $b=16,25$ $c=6,63$	Pcmm	12	1,164 1,177 1,193 1,213 1,164 1,166 1,160 1,147	siehe Abb. 11	8 verzerrt quadratisch antiprismat.	N	7	(39)
$\beta\text{-}Pb(N_3)_2$	$a=18,49$ $b=8,84$ $c=5,12$ $\beta=107,6$	C2 (Cm, C2/m?)	8?	?	?	?	X	—	(40)

Abstände hat (1.21 und 1.15 Å). Letztere ist von den vier Pb-Atomen derart umgeben, daß das endständige N-Atom mit der kürzeren N—N-Bindung nur einem Pb-Atom im Abstand von 2,67 Å benachbart ist, während das andere Azidendatom drei Pb-Nachbarn im Abstand von $1 \times 2,61$ und $2 \times 2,90$ Å hat (Abb. 11). Jedes der endständigen N-Atome der zwei symmetrischen und der einen wenig asymmetrischen N_3-Gruppe steht im Kontakt mit zwei Pb-Atomen mit Pb-N-Abständen zwischen 2.58 und 2.90 Å. Jedes Bleiatom ist von acht N-Atomen in einer verzerrt tetragonal antiprismatischen Anordnung umgeben.

7. Molekulare Azide

Da die molekularen Azide fast durchweg so aufgebaut sind, daß nur eines der N-Atome der Azidogruppe an andere Atome gebunden ist, sind sie von hervorragender Bedeutung für die präparative Chemie. Dieser Aufbau begünstigt nämlich die Abspaltung eines N_2-Moleküls, wobei ein hochreaktives Nitren zurückbleibt, welchem eine Vielzahl von Reaktionspartnern angeboten werden können oder das sich durch Umlagerung stabilisiert. Außerdem können mit kovalent gebundenen Azidogruppen 1,3-dipolare Additionsreaktionen ausgeführt werden, wodurch sich stickstoffreiche Heterocyclen darstellen lassen. Die gute Löslichkeit vieler molekularer Azide in unpolaren Lösungsmitteln erleichtert ihren Einsatz bei chemischen Reaktionen und vermindert dabei gleichzeitig die Explosionsgefahr.

Tabelle 3 gibt einen Überblick über bekannte Strukturdaten von molekularen Aziden. Hierzu gehören die Azide der Nichtmetalle und Halbmetalle sowie Übergangsmetallkomplexe. Bei letzteren unterscheiden sich die beiden N—N-Bindungslängen innerhalb der Azidogruppe im allgemeinen etwas weniger als bei den Nichtmetallaziden.

Die Analyse der thermischen Schwingung zeigt bei molekularen Aziden, deren N_3-Gruppe nur mit dem α-N-Atom gebunden ist, daß die Schwingungsamplitude vom α-N zum γ-N-Atom hin zunimmt; in erster Näherung kann daher die N_3-Gruppe als einigermaßen starres Gebilde angesehen werden, dessen thermische Schwingung sich vor allem in einer Deformation des Bindungswinkels am α-N-Atom äußert. Die mittlere Schwingungsamplitude des γ-N-Atoms kann bei Zimmertemperatur über 0,3 Å liegen.

Die einfachsten molekularen Azide mit bekannter Struktur sind die Verbindungen HN_3 (43, 44), CH_3N_3 (45, 46) und ClN_3 (47). Alle drei haben praktisch denselben Aufbau (Abb. 12) wenn auch auffällt, daß das ClN_3 im Gegensatz zu den anderen beiden eine leicht geknickte

Azidogruppe hat. Wegen ihres einfachen Aufbaus sind CH_3N_3 und vor allem HN_3 die Musterobjekte für theoretische Berechnungen und Betrachtungen bezüglich der Bindungsverhältnisse kovalent gebundener Azidogruppen (vgl. Abschnitt 4).

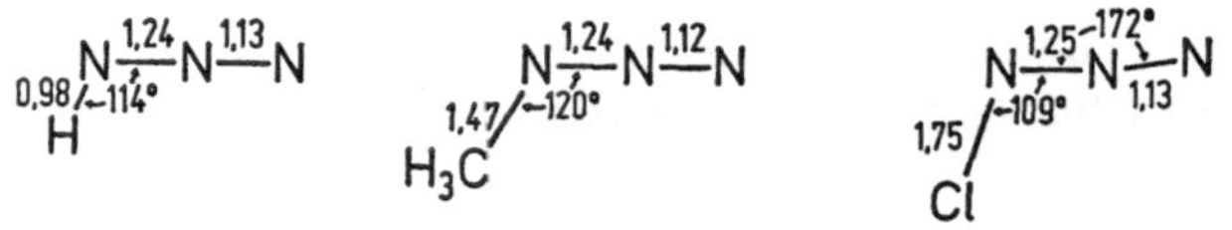

Abb. 12. Struktur der Moleküle HN_3, CH_3N_3 und ClN_3

Azide des Kohlenstoffs, bei denen die Azidogruppe einem π-Bindungssystem benachbart ist, haben die unerwartete Eigenschaft, daß die N_3-Gruppe einen leichten Knick hat und gegen die Ebene des π-Systems geneigt ist (Abb. 13). Der Neigungswinkel beträgt 2°, 3° und 8° beim Triazido-carbonium-ion $[C(N_3)_3]^+$ (49), 15° beim Azido-diamino-carbonium-ion $[N_3C(NH_2)_2]^+$ (50), 8° beim 3-Azido-tropon (69), 5° beim p-Azido-nitrobenzol (51) und 20° beim 1-Azido-2.4.6-trinitrobenzol (52). Bei letzterem hängt der große Neigungswinkel und vielleicht auch der Knick großenteils mit der sterischen Behinderung durch die ortho-ständigen Nitrogruppen zusammen; bei den anderen Verbindungen gibt es keine klare Erklärung für diese Erscheinung, denn die mögliche Wechselwirkung zwischen den π-Bindungssystemen der Azidogruppe und des restlichen Moleküls sollte einen völlig planaren Aufbau begünstigen. Nur beim Cyanurtriazid scheinen die Azidogruppen in der Ebene des Cyanurringes zu liegen (48, 75). Das Triazidocarbonium-ion fällt außerdem dadurch auf, daß es das einzige bekannte Azid ist, bei welchem die N—N-Abstände in den Azidogruppen so verschieden sind (im Mittel 1,39 und 1,05 Å), daß sie eindeutig als Einfach- und Dreifachbindungen anzusehen sind. Dies hängt damit zusammen, daß die p-Orbitale der α-N-Atome am π-Bindungssystem des guanidiniumartigen $C(\alpha N)_3$-Grundgerüsts beteiligt sind und offenbar nicht mehr für das Bindungssystem innerhalb der Azidogruppen zur Verfügung stehen. Dies gilt nicht für das $[N_3C(NH_2)_2]^+$-Ion, dessen N—N-Bindungslängen im Gegensatz zu denen des $[C(N_3)_3]^+$-Ions völlig denen anderer kovalent gebundener Azidogruppen entsprechen. Die theoretische Vorstellung, daß auch in kovalent gebundenen Azidogruppen das mittlere N-Atom eine positive Partialladung trägt (siehe Abschnitt 4), wird durch die Packungsverhältnisse im kristallinen $[C(N_3)_3]^+$ $SbCl_6^-$ dadurch bestätigt, daß jeweils drei Cl-Atome des Anions an die β-N-Atome des Kations anliegen.

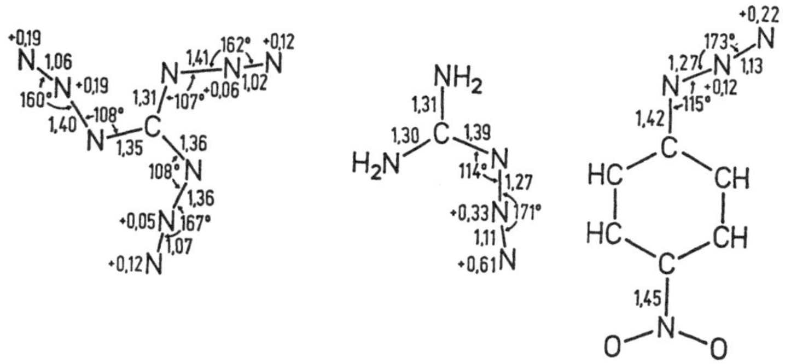

Abb. 13. Struktur der Ionen $[C(N_3)_3]^+$ und $[(H_2N)_2CN_3]^+$ und des Moleküls p-Nitrophenylazid. Zahlen mit Vorzeichen geben an, wie viele Å die betreffenden Atome über der Ebene durch das planare Ionen- bzw. Molekülgrundgerüst liegen

In den Verbindungen $(BCl_2N_3)_3$ (*53*), $(SbCl_4N_3)_2$ (*54*) und $(TaCl_4N_3)_2$ (*55*) sind die α-N-Atome der Azidogruppen an jeweils zwei weitere Atome gebunden und es kommt zur Ausbildung von Sechs- bzw. Vierringen (Abb. 14). Die N—N-Bindungslängen unterscheiden sich nicht signifikant von denen in einfach kovalent gebundenen Aziden, woraus zu schließen ist, daß es für die Bindungen innerhalb der Azidogruppe keine Rolle spielt, ob das α-N-Atom ein freies Elektronenpaar trägt oder stattdessen eine Bindung betätigt. Das $(B-\alpha N)_3$-Ringgerüst des $(BCl_2N_3)_3$ hat die Konformation einer schiefen Wanne (Twist-Form), bedingt durch den Platzbedarf der Cl-Liganden; die Azidogruppen liegen jeweils in einer Ebene mit den an sie gebundenen Boratomen. Dagegen sind beim $(SbCl_4N_3)_2$ und z. T. beim $(TaCl_4N_3)_2$ die Azidogruppen deutlich aus der Ebene des planaren Vierrings herausgewinkelt (beim $(SbCl_4N_3)_2$ 24°), so daß an den α-N-Atomen die Konfiguration einer flachen Pyramide vorliegt; dies ist möglicherweise eine Folge des im Vierring erzwungenen kleinen Sb—N—Sb- bzw. Ta—N—Ta-Winkels von 111° bis 113°, der den α-N-Atomen einen teilweisen sp^3-Hybrid-Charakter aufprägen mag (vgl. auch die Struktur von $Cu(N_3)_2$, Abschnitt 6). Da der Vierring unter Spannung steht, ist das $(SbCl_4N_3)_2$ sehr reaktionsfähig und tauscht seine Azidogruppen leicht gegen Chlor aus. Eine ähnliche Konstitution mit einem Vierring weist auch das Anion im $[As(C_6H_5)_4]_2[(N_3)_2Pd(N_3)_2Pd(N_3)_2]$ auf, bei welchem jedes Pd-Atom an zwei alleinstehende und an zwei Brücken-N_3-Gruppen gebunden ist (*70*). Im Ion $[(CO)_3Mn(N_3)_3Mn(CO)_3]^-$ sind die zwei Mn-Atome über die α-N-Atome der drei Azidogruppen verbrückt (Abb. 15) (*73*).

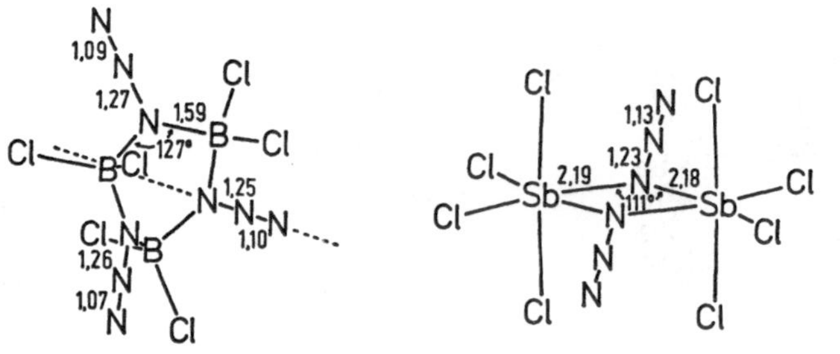

Abb. 14. Struktur der Moleküle $(BCl_2N_3)_3$ und $(SbCl_4N_3)_2$. Das $(BCl_2N_3)_3$ besitzt die Konformation einer schiefen Wanne mit einer gut erfüllten zweizähligen Achse (gestrichelt angedeutet). Das $(SbCl_4N_3)_2$-Molekül ist streng zentrosymmetrisch. Analog wie das $(SbCl_4N_3)_2$ ist auch das $(TaCl_4N_3)_2$ aufgebaut

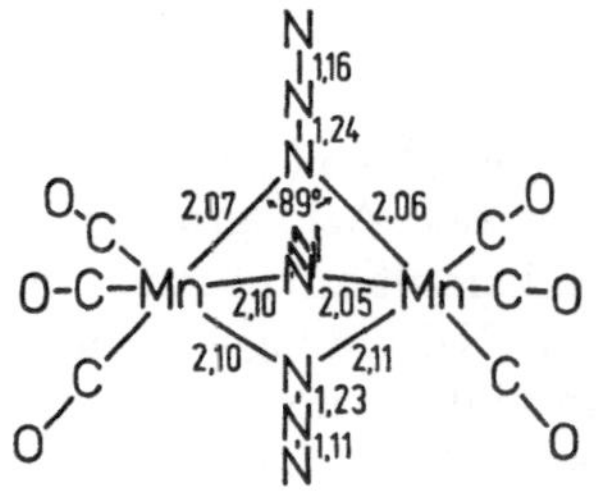

Abb. 15. Struktur des Ions $[(OC)_3Mn(N_3)_3Mn(CO)_3]^-$

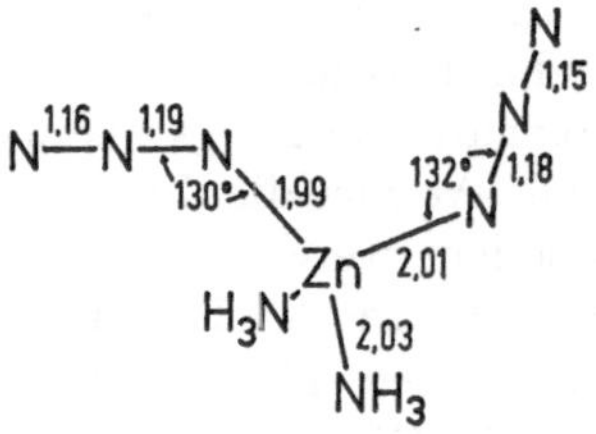

Abb. 16. Struktur eines Moleküls $Zn(NH_3)_2(N_3)_2$. Im Kristallverband ist ein weiteres Molekül vorhanden, das sich nur wenig von dem abgebildeten unterscheidet

Von den Aziden des Zinks wurden die Verbindungen $Zn(N_3)_2(NH_3)_2$ (*56*) und $Zn(N_3)_2(Pyridin)_2$ (*57*) untersucht, in welchen das Zink tetraedrisch koordiniert ist (Abb. 16). Ähnlich liegen die Verhältnisse beim

Komplex $Ni(NO)(P(C_6H_5)_3)_2N_3$, dessen Ni-Atom pseudotetraedrisch von zwei Phosphor- und zwei Stickstoffatomen umgeben ist (58). Pseudotetraedrisch ist auch das Kupfer in der Verbindung $Cu_2(N_3)_2$ [$(CH_2P(C_6H_5)_2)_2$]$_3$ umgeben; jedes Cu ist an drei Phosphinliganden und eine Azidogruppe gebunden (42).

Azidokomplexe mit fünffach koordiniertem Metallatom sind vom Kupfer und Eisen bekannt. Beim Kupfer-(tetraäthyl-diäthylentriamin)-bromid-azid $Cu(Et_4trien)BrN_3$ besetzt das Brom eine äquatoriale und die Azidogruppe eine achsiale Position einer trigonalen Bipyramide, und die Amin-N-Atome die restlichen drei Koordinationsstellen (59). In der Verbindung [$As(C_6H_5)_4$]$_2$[$Fe(N_3)_5$] ist das Eisen in trigonal bipyramidaler Anordnung von den α-N-Atomen der fünf Azidogruppen umgeben (60). Für vier der N_3-Gruppen beträgt der Fe$-$N$-$N-Bindungswinkel 125°, während er bei der fünften 146° betragen soll; diese N_3-Gruppe soll in ihrer Orientierung nicht festliegen. Die Fe$-$N-Abstände betragen 2,04 Å für die achsialen und 1,97 Å für die äquatorialen Liganden; für alle N$-$N-Bindungen wird der gleiche Abstand von 1,16 Å angegeben.

Oktaedrisch koordinierte Metallatome findet man bei den Amino-Azido-Komplexen des Kobalts und des Rutheniums. Beim (4-(2-Amino-äthyl)-1.4.7.10-tetrazadecano)-azido-kobalt-nitrat besetzen die Aminogruppen des Chelatliganden fünf Koordinationsstellen und die Azidogruppe die sechste (61); alle sechs Co$-$N-Abstände sind fast gleich. Die Azidogruppe hat ekliptische Konformation zu einem der Aminoliganden (Abb. 17). Auch beim Bisäthylendiamino-diazido-kobalt-nitrat [Co en$_2$(N$_3$)$_2$]NO$_3$ sind alle sechs Co$-$N-Abstände annähernd gleich (62). Die zwei Azidogruppen nehmen cis-Konfiguration ein, sind linear, unterscheiden sich aber etwas in ihren N$-$N-Bindungsabständen. Die βN$-\gamma$N-Abstände sollen dabei größer sein als die αN$-\beta$N-Abstände, was im Widerspruch zu den anderen Azidstrukturen steht; vermutlich liegt hier ein Fehler in den Angaben vor. Zur gleichen Verbindungsklasse ist das Bisäthylendiamino-nitrogeno-azido-ruthenium-hexafluorophosphat [Ru en$_2$(N$_2$)N$_3$] PF$_6$ zu zählen (64). Hier ist einer der sechs Liganden ein N_2-Molekül, das sich in trans-Position zur Azidogruppe befindet (Abb. 17). Der Ru$-$N-Abstand zu diesem N_2-Liganden ist mit 1.89 Å deutlich kürzer als die übrigen Ru$-$N-Abstände, so daß hier eine stärkere Metall-Stickstoff Wechselwirkung angenommen werden muß. Ein Beispiel für eine Verbindung, die gleichzeitig der Gruppe der molekularen und der ionischen Azide angehört, ist [$Co(NH_3)_5N_3$]$^{++}$(N$_3^-$)$_2$ (63). Hier liegt ein komplexes Kation mit einer kovalent gebundenen Azidogruppe vor und die Anionen sind praktisch symmetrische Azidionen.

Eines der seltenen molekularen Azide, bei welchem beide endständigen N-Atome der Azidogruppe je eine kovalente Bindung betätigen, ist die

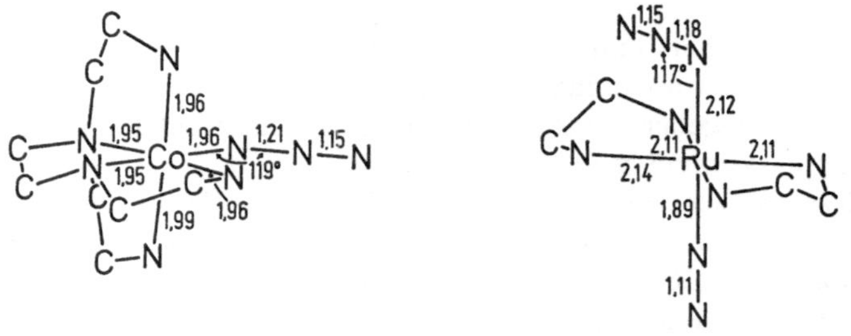

Abb. 17. Struktur der Chelatkomplexe [Co(H₂NC₂H₄NHC₂H₄N(C₂H₄NH₂)₂)N₃]⁺⁺ und [Ru(N₂)(H₂NC₂H₄NH₂)₂N₃]⁺. Zur besseren Übersicht wurden die H-Atome weggelassen

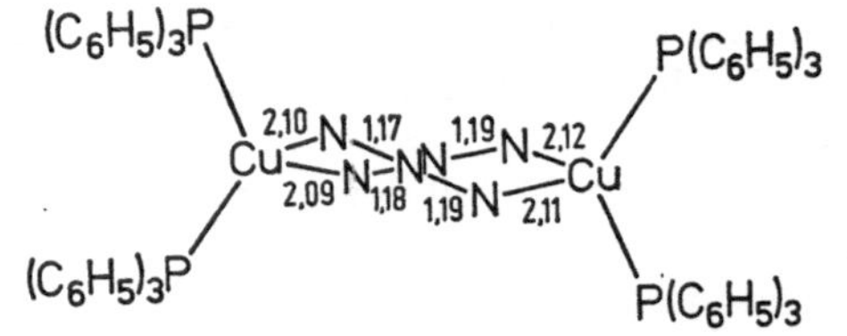

Abb. 18. Struktur der Verbindung [Cu(P(C₆H₅)₃)₂N₃]₂. Die Azidogruppen sind mit beiden endständigen N-Atomen an je eines der Cu-Atome gebunden unter Ausbildung eines Achtringes

Verbindung [Cu(P(C₆H₅)₃)₂N₃]₂ (*65*). Die beiden pseudotetraedrisch umgebenen Cu-Atome sind über zwei in sich praktisch symmetrische N₃-Gruppen miteinander verbunden. Bei jeder Azidogruppe stehen die N—Cu-Bindungen in einer gauche Konformation zueinander, so daß eine etwas dem Allen entsprechende Konfiguration zu erkennen ist (Abb. 18) (vgl. hierzu die Struktur von Cd(N₃)₂(Pyridin)₂, Abschnitt 6).

8. Ergänzung

Während der Drucklegung dieser Übersicht wurden die Strukturen der folgenden Azide publiziert: CH₃N₃ und SiH₃N₃ (*76*), GeH₃N₃ (*77*), NCN₃ (*78*), K[(CH₃)₃AlN₃Al(CH₃)₃] (*79*), Cu(HN((CH₂)₂N(C₂H₅)₂)₂ BrN₃ (*80*), CH₃HgN₃ (*81*), α-Hg(N₃)₂ (*82*).

Tabelle 3. *Übersicht über die wichtigsten Strukturdaten von molekularen Aziden*
Es gelten die Angaben wie in der Legende zu Tabelle 1. Bestimmungsmethoden: X = Röntgen., E = Elektronenbeugung, M = Mikrowellenspektroskopie. en = 1.2-Diaminoäthan, Ph = Phenyl, Et = Äthyl, Py = Pyridin

Verbindung	Elementarzelle	Raumgruppe	Formeleinh. pro El.-zelle	Abstände $\alpha N-\beta N$	$\beta N-\gamma N$	$R-\alpha N$	Winkel $R-\alpha N-\beta N$	Best.-methode	R(%)	Literatur
HN_3	—	—	—	1,237	1,133	0,98	114,1	M	—	(43,44)
$(BCl_2N_3)_3$	$a=8,874\ b=14,494$ $c=10,538\ \beta=99,74$	$P2_1/c$	4	1,27 1,26 1,25	1,09 1,07 1,10	1,55 bis 1,60	113,2 bis 118,1	X	5,8	(53)
CH_3N_3	—	—	—	1,24	1,12	1,47	120	E	—	(45,46)
$CH_3N_3 \cdot SbCl_5$	$a=7,03\ b=23,73$ $c=12,13\ \beta=90,3$	$P2_1/c$	8	?	?	?	?	X	—	(68)
$(CH_3)_2C$ (siehe Struktur)	$a=5,86\ b=10,20$ $c=10,80\ \beta=91,2$	$P2_1$	2	1,08[a]	1,20[a]	1,62[a]	114	X	19,7	(66)
Ph_3CN_3	$a=8,85\ b=15,42$ $c=17,17\ \alpha=139,6$ $\beta=91,8\ \gamma=91,6$	$P\bar{1}$	4	?	?	?	?	X	—	(68)
3-Azidotropon	$a=14,280\ b=12,786$ $c=7,757$	Pbca	8	1,25	1,12	1,42	116	X	6,5	(69)
p-Nitrophenylazid	$a=18,05\ b=10,29$ $c=3,73$	$P2_12_12_1$	4	1,27	1,13	1,42	115,0	X	8,7	(51)

Tabelle 3 (Fortsetzung)

Verbindung	Elementarzelle	Raumgruppe	Formeleinh. pro El.-zelle	Abstände αN–βN	βN–γN	R–αN	Winkel R–αN–βN	Best.-methode	R(%)	Literatur
[2.4.6-Trinitrophenylazid]₂·Cu(hydroxychinolin)₂	$a=16{,}14\ b=30{,}93$ $c=6{,}90\ \beta=105{,}6$	A2/a	4	1,24	1,12	1,46	119,2	X	13,4	(52)
Cyanurtriazid	$a=8{,}75\ c=5{,}85$ (bei −110 °C)	P6₃/m (P6₃?)	2	1,27	1,12	1,40	112,2	X	8,1	(48, 75)
[(NH₂)₂CN₃]Cl	$a=5{,}116\ b=10{,}737$ $c=9{,}707\ \beta=92{,}2$	Cc	4	1,27	1,11	1,39	114,2	X	3,2	(50)
[NH₂C(N₃)₂]SbCl₆	$a=19{,}73\ b=11{,}02$ $c=12{,}17$	Pnma (Pn2₁a?)	8	?	?	?	?	X	—	(67)
[C(N₃)₃] SbCl₆	$a=9{,}27\ b=10{,}96$ $c=15{,}30\ \beta=113{,}4$	P2₁/c	4	1,41 1,36 1,40	1,02 1,07 1,06	1,31 1,36 1,35	107,3 108,0 108,0	X	6,8	(49)
(SbCl₄N₃)₂	$a=8{,}05\ b=9{,}353$ $c=10{,}12\ \beta=93{,}8$	P2₁/n	2	1,23	1,13	2,18 2,19	119,6 122,3	X	4,9	(54)
ClN₃	—	—	—	1,252	1,133	1,745	108,7	M	—	(47)
[Cu(PPh₃)₂N₃]₂	$a=23{,}524\ b=13{,}690$ $c=20{,}035\ \beta=106{,}3$	P2₁/c	4	1,17 1,18	1,19 1,19	2,10 2,12 2,09 2,11	122,1 118,1 124,4 124,6	X	6,5	(65)

$Cu_2[(CH_2PPh_2)_2]_3(N_3)_2$	$a = 18,06\ b = 18,63$ $c = 21,02$	Pbca	4	1,20	1,08	2,04	132	X	6,5	(42)
$Cu[NH(C_2H_4NEt_2)_2]BrN_3$	$a = 12,95\ b = 7,65$ $c = 9,87\ \alpha = 80,7$ $\beta = 113,2\ \gamma = 100,0$	P$\bar{1}$	2	nicht angegeben				X	10	(59)
$Zn(NH_3)_2(N_3)_2$	$a = 9,565\ b = 7,158$ $c = 18,976$	Pnma	8	{1,19 / 1,18} {1,20 / 1,17}	1,16 / 1,15 1,14 / 1,18	1,99 / 2,01 1,99 / 1,97	130 / 132 126 / 126	X	7,6	(56)
$ZnPy_2(N_3)_2$	$a = 8,528\ b = 20,161$ $c = 8,064\ \beta = 106,1$	P2$_1$/c	4	1,17 / 1,15	1,13 / 1,13	1,93 / 1,95	129 / 129	X	7,5	(57)
$K_2Zn(N_3)_4$	$a = 14,40\ b = 11,70$ $c = 11,29$	Pbca	8	?	?	?	?	X	—	(41)
$(TaCl_4N_3)_2$	$a = 13,662\ b = 8,821$ $c = 13,188\ \beta = 112,1$	P2$_1$/c	4	1,18 1,25	1,15 1,16	2,14 / 2,21 / 2,18 / 2,18	125 / 118 / 124 / 114	X	8,2	(55)
NEt_4 $[(CO)_3Mn(N_3)_3Mn(CO)_3]$	$a = 10,30\ b = 10,21$ $c = 21,92\ \beta = 91,1$	P2$_1$/n	4	1,24 / 1,23 / 1,20	1,16 / 1,11 / 1,20	2,05 bis 2,11	125	X	10	(73)
$(AsPh_4)_2[Fe(N_3)_5]$	nicht angegeben	C2/c	4	1,16	1,16	1,96 bis 2,04	4 × 125 1 × 146?	X	9	(60)
$[Ru(N_2)en_2N_3]\,PF_6$	$a = 9,97\ b = 12,01$ $c = 12,59\ \beta = 102,4$	P2$_1$/n	4	1,18	1,15	2,12	116,7	X	5,6	(64)
$[Co(NH_3)_5N_3](N_3)_2$	$a = 12,997\ b = 8,031$ $c = 10,414$	Pnam	4	1,21	1,15	1,94	125,2	X	7,5	(63)

Tabelle 3 (Fortsetzung)

Verbindung	Elementarzelle	Raum-gruppe	For-mel-einh. pro El.-zelle	Abstände αN–βN	βN–γN	R–αN	Winkel R–αN–βN	Best.-me-thode	R(%)	Lite-ratur
[Co(NH$_2$C$_2$H$_4$NHC$_2$H$_4$N(C$_2$H$_4$NH$_2$)$_2$)N$_3$](NO$_3$)$_2$ · H$_2$O	$a=8{,}32$ $b=7{,}64$ $c=27{,}69$ $\beta=96{,}3$	P2$_1$/c	4	1,21	1,15	1,96	119,0	X	6,8	(61)
[Coen$_2$(N$_3$)$_2$] NO$_3$	$a=12{,}106$ $b=23{,}62$ $c=8{,}801$	Pnma	8	1,14$^b)$ 1,11$^b)$	1,23$^b)$ 1,14$^b)$	1,95 1,97	119 120	X	10	(62)
NiNO(PPh$_3$)$_2$N$_3$	$a=13{,}691$ $b=19{,}211$ $c=12{,}582$ $\beta=98{,}1$	P2$_1$/c	4	0,98$^c)$	1,28$^c)$	2,02	128,1	X	5,2	(58)
(AsPh$_4$)$_2$[Pd$_2$(N$_3$)$_6$]	$a=11{,}43$ $b=11{,}78$ $c=10{,}36$ $\alpha=109{,}0$ $\beta=103{,}1$ $\gamma=100{,}4$	P$\bar{1}$	2	1,22 1,20 1,24	1,12 1,16 1,14	1,99 2,01	122,1 119,2 128,4 127,5	X	5,8	(70)

N–N-Abstände, die nicht im Einklang mit den übrigen Strukturen stehen;
a) Diese Werte sind wahrscheinlich sehr ungenau, wie der relativ hohe R-Wert verrät.
b) Bei dieser Strukturbestimmung wurden 85 Parameter mit nur 293 Reflexintensitäten ermittelt, so daß die Werte ziemlich ungenau sein müssen.
c) Der Autor gibt diese Abstände als unbefriedigend an; es wird für möglich gehalten, daß der untersuchte Kristall noch Chlor an Stelle von Azid enthielt.

9. Literatur

Zusammenfassende Darstellungen:
1. *Boyer, H. J., Canter, F. C.:* Chem. Rev. *54, 1* (1954).
2. *Grundmann, C.:* In: Houben-Weyl, Methoden der organischen Chemie; Bd. 10/3, S. 777. Stuttgart: G. Thieme Verlag 1965.
3. *Lieber, E., Minnis, R. L., Rao, C. N. R.:* Chem. Rev. *65*, 377 (1965).
4. *— Curtice, J. S., Rao, C. N. R.:* Chem. Ind. *1966*, 586.
5. *Lappert, M. F., Pyszora, H.:* Advan. Inorg. Chem. Radiochem. *9*, 133 (1966).
5a.*Thayer, J. S.:* Organometal. Chem. Rev. *1*, 157 (1966).
6. *Dehnicke, K.:* Angew. Chem. *79*, 253 (1967).
6a.*Paetzold, P.:* Fortschr. Chem. Forsch. *8*, 437 (1967).
7. *L'abbé, G.:* Chem. Rev. *69*, 345 (1969).
8. *— Hassner, A.:* Angew. Chem. *83*, 103 (1971).
9. *Patai, S.* (Herausgeber): The Chemistry of the azido group. London: Interscience Publishers 1971.
10. *Evans, B. L., Yoffe, A. D., Gray, P.:* Chem. Rev. *59*, 515 (1959).
11. *Bowden, F. P., Yoffe, A. D.:* Fast Reactions in Solids. London: Butterworths Scientific Publications 1958.
12. *Gray, P.:* Quart. Rev. *17*, 441 (1963).
12a.*Iqbal, Z.:* Struct. Bonding *10*, 25 (1972).
13. *Yoffe, A. D.:* Develop. Inorg. Nitrogen Chem. *1*, 72 (1966).
14. *Wyckoff, R. W. G.:* Crystal structures; 2nd edition, Vol. 2, p. 278—283. New York: Interscience Publishers, 1964.
15. *Waddington, T. C.:* Advan. Inorg. Chem. Radiochem. *1*, 157 (1959).
16. *Gillespie, R. J.:* Angew. Chem. *79*, 885 (1967).
17. *Pauling, L.:* Die Natur der chemischen Bindung,; S. 246. Weinheim: Verlag Chemie 1962.

Einzelpublikationen:
Nur die neuesten Arbeiten werden zitiert, sofern sich darin Hinweise auf ältere Arbeiten finden.
18. *Wyatt, J. F., Hillier, I. H., Saunders, U. R., Connor, J. A., Barber, M.:* J. Chem. Phys. *54*, 5311 (1971).
19. *Pringle, G. E., Noakes, D. E.:* Acta Cryst. B *24*, 262 (1968).
20. *Müller, U.:* Z. Anorg. Allgem. Chem., *392*, 159 (1972).
21. *Mueller, H. J., Joebstl, J. A.:* Z. Krist. *121*, 385 (1965).
22. *Frevel, L. K.:* Z. Krist. *94*, 197 (1936).
23. *Boutin, H., Trevino, S., Prask, H.:* J. Chem. Phys. *45*, 401 (1966).
24. *Krischner, H.:* Monatsh. Chem. *99*, 2134 (1968).
25. *Choi, C. S.:* Acta Cryst. B *25*, 2638 (1969).
26. *Walitzi, E. M., Krischner, H.:* Z. Krist. *132*, 19 (1970).
27. *Griffin, J. F., Coppens, P.:* Chem. Commun. *1971*, 502.
28. *Walitzi, E. M., Krischner, H.:* Z. Krist. *131*, 25 (1970).
29. *Marr, H. E., Stanford, R. H.:* Acta Cryst. *15*, 1313 (1962).
30. *Bassière, M.:* Compt. Rend. *201*, 735 (1935).
31. *West, C. D.:* Z. Krist. *95*, 42 (1936).
32. *Wilsdorf, H.:* Acta Cryst. *1*, 115 (1948).
33. *Agrell, I., Lamnevik, S.:* Acta Chem. Scand. *22*, 2038 (1968).
34. *Söderquist, R.:* Acta Cryst. B *24*, 450 (1968).
35. *Agrell, I.:* Acta Chem. Scand. *20*, 1281 (1966).
36. *—* Acta Chem. Scand. *23*, 1667 (1969).
37. *Bassière, M.:* Compt. Rend. *204*, 1573 (1937).

38. *Agrell, I.:* Acta Chem. Scand. *24,* 3575 (1970).
39. *Choi, C. S., Boutin, H. P.:* Acta Cryst. *B 25,* 982 (1969).
40. *Azdroff, L. V.:* Z. Krist. *107,* 362 (1956).
41. *Krischner, H., Fritzer, H. P.:* Z. Anorgan. Allgem. Chem. *376,* 162 (1970).
42. *Gaughan, A. P., Ziolo, R. F., Dori, Z.:* Inorg. Chem. *10,* 2776 (1971).
43. *Amble, E., Dailey, B. P.:* J. Chem. Phys. *18,* 1437 (1950).
44. *Winnewisser, M., Cook, R. L.:* J. Chem. Phys. *41,* 999 (1964).
45. *Pauling, L., Brockway, L. O.:* J. Am. Chem. Soc. *59,* 13 (1937).
46. *Livingston, R. L., Rao, C. N. R.:* J. Chem. Phys. *64,* 756 (1960).
47. *Cook, R. L., Gerry, M. C. L.:* J. Chem. Phys. *53,* 2525 (1970).
48. *Knaggs, I. E.:* Proc. Roy, Soc. (London) *A 150,* 576 (1935).
49. *Müller, U., Bärnighausen, H.:* Acta Cryst. *B 26,* 1671 (1970).
50. *Henke, H., Bärnighausen, H.:* Acta Cryst., *B 28,* 1 100 (1972).
51. *Mugnoli, A., Mariani, C., Simonetta, M.:* Acta Cryst. *19,* 367 (1965).
52. *Bailey, A. S., Prout, C. K.:* J. Chem. Soc. *1965,* 4867.
53. *Müller, U.:* Z. Anorg. Allgem. Chem. *382,* 110 (1971).
54. — Z. Anorg. Allgem. Chem., *388,* 207 (1972).
55. *Strähle, J.:* Z. Anorg. Allgem. Chem., im Druck.
56. *Agrell, I., Vannerberg, N. G.:* Acta Chem. Scand. *25,* 1630 (1971).
57. — Acta Chem. Scand. *24,* 1247 (1970).
58. *Ennemark, J. H.:* Inorg. Chem. *10,* 1952 (1971).
59. *Dori, Z.:* Chem. Commun. *1968,* 714.
60. *Drummond, J., Wood. J. S.:* Chem. Commun. *1969,* 1373.
61. *Maxwell, I. E.:* Inorg. Chem. *10,* 1782 (1971).
62. *Padmanabhan, V. M., Balasubramanian, R., Muralidharan, K. V.:* Acta Cryst. *B 24,* 1638 (1968).
63. *Palenik, G. J.:* Acta Cryst. *17,* 360 (1964).
64. *Davis, B. R., Ibers, J. A.:* Inorg. Chem. *9,* 2768 (1970).
65. *Ziolo, R. F., Gaughan, A. R., Dori, Z., Pierpont, C. D., Eisenberg, R.:* Inorg. Chem. *10,* 1289 (1971).
66. *Brimacombe, J. S., Bryan, J. G. H., Hamor, T. A.:* J. Chem. Soc. *B 1970,* 514.
67. *Buschmann, E., Henke, H., Bärnighausen, H.:* unveröffentlicht.
68. *Henke, H.:* unveröffentlicht.
69. *Cruickshank, D. W. J., Filippini, G., Mills, O. S.:* Chem. Commun. *1972,* 101.
70. *Fehlhammer, W. P., Dahl, L. F.:* J. Am. Chem. Soc., *94,* 3 377 (1972).
71. *Torkar, K., Krischner, H., Radl, H.:* Monatsh. Chem. *96,* 932 (1965).
72. *Dixon, H. P., Jenkins, H. D. B., Waddington, T. C.:* Chem. Phys. Letters *10,* 600 (1971).
73. *Mason, R., Rusholme, G. A., Beck, W., Engelmann, A., Joos, K., Lindenberg, B., Smedal, H. S.:* Chem. Commun. *1971,* 496.
74. *Torkar, K., Krischner, H., Ernst, G.:* Monatsh. Chem. *100,* 203 (1969).
75. *Beineke, T. A.:* Dissertation 1966. University Microfilms (Ann Arbor) 66 — 3. Dissert. Abstracts *26,* 4257 (1966).
76. *Anderson, D. W. W., Rankin, D. W. H., Robertson, A.:* J. Molec. Struct. *14,* 385 (1972).
77. *Murdoch, J. D., Rankin, D. W. H.:* Chem. Commun. *1972,* 748.
78. *Costain, C. C., Kroto, H. W.:* Canad. J. Phys. *50,* 1453 (1972).
79. *Atwood, J. L., Newberry, W. R.:* J. Organometal. Chem. *42,* C77 (1972).
80. *Ziolo, R. F., Allen, M., Titus, D. D., Gray, H. B., Dori, Z.:* Inorg. Chem. *11,* 3044 (1972).
81. *Müller, U.:* Z. Naturforsch., im Druck.
82. — Z. Anorg. Allgem. Chem., im Druck.

Eingegangen am 23. Februar 1972

Structure and Bonding: Index Volume 1-14